Everything is ecology

Mark Maraun

Everything is ecology

palgrave
macmillan

Mark Maraun
Tierökologie / Zoologie Universität
Göttingen
Göttingen, Germany

ISBN 978-3-658-50206-5 ISBN 978-3-658-50207-2 (eBook)
https://doi.org/10.1007/978-3-658-50207-2

This book is a translation of the original German edition "Alles ist Ökologie" by Mark Maraun, published by Springer Fachmedien Wiesbaden GmbH in 2025. The translation was done with the help of an artificial intelligence machine translation tool. A subsequent human revision was done primarily in terms of content, so that the book will read stylistically differently from a conventional translation. Springer Nature works continuously to further the development of tools for the production of books and on the related technologies to support the authors.

This Springer VS imprint is published by the registered company Springer Fachmedien Wiesbaden GmbH, part of Springer Nature.
The registered company address is: Abraham-Lincoln-Str. 46, 65189 Wiesbaden, Germany

For my parents, my wife Patricia, and my children;
the most important resources of my life

Preface

Nowadays, everything is somehow "eco." In retail, there are private labels called "Go Eco"; there is an EU Eco Regulation, there are eco-standards, there is the idea of eco instead of ego, there are eco-villages and eco-cars, eco-houses, green electricity, ecological industry, eco-farming, and eco-balances; ultimately, everything is somehow "eco." Everything and everyone is somehow ecological, but many people don't even know what "eco" actually means. Most of the time, "eco" has a positive connotation. Only conservatives and liberals are somewhat skeptical, because they think that not everyone has to be eco and that there is also a lot of paternalism from all the ecos; sometimes this is referred to as eco-terrorism. Famous ecos, of course, include Greta Thunberg and Luisa Neubauer in Germany. They are committed to preventing or at least limiting climate change so that we can continue to inhabit this planet for many generations to come. Why this can sometimes be a problem will be discussed later.

Often, "eco" is also synonymous with "organic," especially when it comes to food. This may seem somewhat trivial, since the opposite of organic food would be non-organic, such as plastic or concrete, and we are generally not keen on eating those. The opposite of "eco" is somewhat easier to define. In food, agriculture, and forestry, the opposite of "eco" usually involves the addition of pesticides and fertilizers; in the tropics, it refers to the exploitation of forests and, more generally, the destruction of humanity's common goods. "Eco" actually comes from the Greek (from "oikos") and simply means "house," and thus refers to the household of nature; just as "oikos" in the word "economy" simply means household management in economic life. Ecology itself is, at first, value-neutral. It is only in nature conservation that we evaluate the household of nature.

Actually, nowadays "eco" usually means something else, namely sustainability. Sustainability is easier to define. It concerns any form of human economic activity. We (sometimes) strive not to extract more resources from the environment than can be regenerated, for example in sustainable forestry. This brings us to one of humanity's greatest problems: individual interests are generally not identical to group interests. This means that even if I actually consider preserving the climate important, I find it even more important that I and my children have a roof over our heads and enough to eat. Perhaps I also find it more important to be able to afford children at all. What good does it do me as an individual if the climate is saved, but I myself lack the resources to support my family and children? We will return to this topic. Whether and how this problem can be solved will also be addressed in the chapter on the "tragedy of the commons."

For as long as I can remember, I have always wanted to know and understand everything ("Indeed, I know much, yet I wish to know everything," Johann Wolfgang von Goethe has a student say in "Faust"). I wanted to know why things on this earth cannot be more peaceful and just. Why are there so many conflicts, disputes, wars (in 2025/2026 more relevant than ever)? Faust himself later says, "Alas! I have studied philosophy, law, and medicine, and unfortunately also theology, thoroughly, with passionate effort. Here I stand, poor fool that I am! And am as wise as before." After finishing school, I too wanted to know everything, and I wanted at least to get a bit further than Faust. After all, Faust's time was long ago. Perhaps by now, people knew more....

I then started studying philosophy, anthropology, and biology in Göttingen in 1987; indeed, with great enthusiasm... but after a short time, I gave up on philosophy. I had the impression that most philosophers had the misfortune of having thought about things before the development of evolutionary theory, and therefore lacked the most important theory for understanding the world. Only the younger philosophers, who were at least somewhat familiar with biology and evolution (for example, Gerhard Vollmer in Germany and D.C. Dennett in the USA), could, in my opinion, contribute something to our understanding of things on this earth.

I had always been interested in the natural sciences and biology, ecology, and evolutionary biology, and at that time I read Konrad Lorenz and everything I could find on the subject of biology. This was also the time when the ecology movement began; the Green Party was founded. I read the first doomsday scenarios in the books by Hoimar von Ditfurth ("So let's plant an apple tree"), the report from the Club of Rome (roughly equivalent today to the reports of the International Panel of Climate Change (IPCC)), and 'Global 2000', the report to American President Carter on overpopulation, nuclear war, and world hunger.

After Göttingen, I spent time in Darmstadt, in Canada (Calgary), in London, and am now back in Göttingen. The ecology lecture there forms the basis of this book. Many ideas on the topic of ecology came from Professor Stefan Scheu;

however, I take responsibility for all the wild conclusions I draw in this book. But without Stefan Scheu, this book would not exist. I also thank my parents, who have always supported me emotionally and financially, allowing me to work and think in peace; my wife Patricia, who has always stood by my side and encouraged me to pursue my ideas and write a book; and my children, who have always provided me with inspiration from real life. And so that no one has to google it: I am a white, older, heterosexual man, married for the second time. I live in a patchwork family and have four children whose future in health, prosperity, security, and freedom is very close to my heart, and I hope that this book can contribute to that.

This is not an ecological textbook, but rather a book that demonstrates whether and what the science of ecology can contribute to current issues in society and politics.

Göttingen
November 2025

Mark Maraun

Contents

About the Author

Photo: Susanne Wintzenburg

Prof. Dr. Mark Maraun teaches ecology, zoology, and evolution at the University of Göttingen. He has conducted research and worked in Göttingen, Calgary (Canada), London (UK), and Darmstadt. His research focuses on the ecology and evolution of soil-dwelling oribatid mites. A particular emphasis of his work is understanding the high proportion of asexual species in the soil. Mark Maraun is married and has four children.

Prologue—Chance and Determinism

On August 27, 1883, a volcano on the island of Krakatau, located between Java and Sumatra in Indonesia, exploded with the force of about 10,000 atomic bombs (Thornton 1996). The explosion could be heard as far as 3000 km away in Australia. Gigantic tsunami waves were generated, reaching as far as South Africa and Alaska. Approximately 36,000 people died; the animal and plant life in the immediate vicinity of the volcano (on the islands of Sertung, Panjang, and on Rakata—the remainder of the island of Krakatau) was almost completely wiped out. The eruption led to a temporary darkening of the atmosphere and caused global temperatures to drop by about 0.5 degrees. After the explosion, the island of Krakatau had almost disappeared; only about a third remained, with the rest having sunk into the Indian Ocean, forming a huge underwater caldera, the collapse crater of the volcano. At the site of the former volcano, due to ongoing volcanic activity, a small hill gradually emerged from the water over the following decades: Anak Krakatau, the child of Krakatau, which grows by about 60 cm per year.

In the short term, this eruption was, of course, a catastrophe for all living beings in the immediate area; in the long term, however, a succession of animals and plants developed on the remnants of Krakatau and the surrounding islands, which would prove instructive for ecologists and evolutionary biologists. The area around Krakatau was ecologically reset to more or less zero; a *tabula rasa* for plants and animals, ready for recolonization.

Comparable volcanic eruptions have occurred on the Greek island of Santorini; however, these eruptions began about a million years ago and continue to this day. The last eruption took place about 3600 years ago. A somewhat more recent eruption is that of Mount St. Helens in the US state of Washington on July 10, 2008.

M. Maraun, *Everything is Ecology*, https://doi.org/10.1007/978-3-658-50207-2_1

At first glance, all these are examples of random events. While it is possible to predict to some extent whether volcanoes will explode or erupt, the exact date and magnitude of the explosion cannot be perfectly predicted, even with modern technology. How do such events shape the course of evolution? Or do they not shape it at all? Are these perhaps singular events without lasting significance for the animal and plant world? How does recolonization occur in areas that are almost completely devoid of animals and plants? Is recolonization predictable and deterministic, meaning that exactly the same organisms return as those that existed on the island before the catastrophe, or does it occur randomly; are the species that colonize the area around a volcano after an eruption more of a hodgepodge of species that were simply the first to reach the island with their propagules, e.g., seeds? Is it perhaps simply the case that the most successful colonizer is the one who arrives first (Hubbel 2001)? These are the kinds of questions addressed by the theory of island biogeography.

Is an event like a volcanic eruption really to be considered random and thus an unpredictable catastrophe for plants? The plants that live in volcanic regions may be well adapted to such a habitat, which is shaped by abiotic catastrophes; at the very least, they can recolonize it quickly. Just as there are plants in regions where fires are frequent. For these species, fire is not a catastrophe at all; some species are even so-called fire-germinators (e.g., the true acacia, *Acacia pulchella*, an evergreen shrub from southwestern Australia), meaning they only germinate after a fire; only the heat allows the seeds to develop. And what about us humans? Do coincidences exist for us, or do we construct our own environment? Do we live exactly where we best fit? Do we create our own "coincidences"? What do the religions of this world say about the concept of "chance"?

About 66 million years ago, a giant meteorite with a diameter of about 10 km struck the Yucatan Peninsula in what is now Mexico. It had the force of about 200 million atomic bombs and left behind the so-called Chicxulub crater, which shortly after the impact was 35 km deep and 200 km in diameter (Goderis et al. 2021). Today, the crater can only be made visible with difficulty and with the help of computer-assisted methods. The existence of this crater was proven by Alvarez and colleagues in the 1970s (Alvarez et al. 1980). They were able to show that rocks from the time of the meteorite impact contained elevated concentrations of iridium (the so-called iridium anomaly). Such high concentrations of iridium are otherwise only found in meteorites and point to a massive cosmic impact scenario (Alvarez 1997). After the impact, there was an impact winter with low temperatures and reduced photosynthesis, followed—due to the high amounts of CO_2 in the atmosphere—by a prolonged warm phase.

The cause of the impact is still unknown. Either it was pure chance (from time to time, meteorites simply cross Earth's path), or there may be a second small sun (sometimes called Nemesis) that, on its orbit around the sun every 26 million years, knocks comets out of the Oort cloud, some of which then strike Earth. The latter theory, however, is purely speculative; so far, there are no data or observations to support it (Raup 1992). For our purposes, however, the cause is of little importance for now.

The impact of this meteorite affected the Earth even more profoundly than the eruption of Krakatau. After the impact, megatsunamis occurred, followed by earthquakes of magnitude 11–12 that were felt all over the world. The shock triggered massive tectonic faults, leading to worldwide volcanic eruptions and large releases of CO_2; the Earth was darkened for several years. Due to this darkening, photosynthesis and thus the primary production of plants almost completely ceased, ultimately leading to the extinction of the dinosaurs as the most famous example. In addition to the dinosaurs, parts of the marine fauna, many terrestrial species (Longrich et al. 2011), and many plant taxa (~60%; Fawcett et al. 2009) also became extinct. Many millions of years later, we humans would call this transition the Cretaceous-Tertiary (KT) boundary, which in 2000 was renamed the Cretaceous–Paleogene (KPg) boundary, because the Tertiary was originally used only to name rock formations and was not actually a time division.

Were all groups of plants and animals equally affected? Apparently not; mammals and their ancestors survived and subsequently underwent an impressive radiation (Halliday et al. 2016), as did most birds (which evolved from small, running dinosaurs) (Cooper and Penny 1997). One group probably suffered little from the impact; on the contrary, they may even have benefited from the catastrophe: the decomposers in the soil, such as woodlice, mites, and springtails (Collembola). They consume dead organic material, i.e., leaves and needles lying on the ground, dead animals, as well as bacteria and fungi that colonize these former living beings. The explosion likely had little effect on them (except at the impact site). No extinction event is known for them that could be detected with modern methods, such as a molecular clock (Schaefer et al. 2010; Pachl et al. 2021).

Chance can literally be fatal for us, for example, if a real (not symbolic) roof tile unfortunately falls on our head. At the very least, chance often leaves us in despair; sometimes it drives us mad. In accidents and terrorist attacks, it can plunge us into despair when relatives and friends are affected. Some misfortunes are so hard to understand (why was my acquaintance/friend/relative on exactly the plane that crashed?) that we can hardly cope. We look for causes, want to find someone to blame, want to understand, instead of accepting that it was simply chance, that, e.g., a large bird flew into the engine. We ask ourselves: why was that particular person in the wrong place at the wrong time? Why were they hiking on the volcano when it erupted? Why was it my friend who was caught in that avalanche? We search for a reason, because we are used to it, so that we can better understand the world. We do not particularly like chance. It has a certain meaninglessness. Chance leads to uncertainty, helplessness, and fear.

However, there are also pleasing positive chance events. Winning the lottery is, of course, very gratifying; finding a life partner as well (even if, at the moment of finding, this is not always entirely clear). And even the conception of the reader was an extremely unlikely and thus random event, resulting from all the possible combinations of the egg cell with one of the millions of sperm. In these cases, despair is usually not as great; nor is the search for meaning as intense. Do we deserve to win the lottery; were we truly ready in our hearts for our dream partner; did we also construct this positive reality for ourselves? And does that mean nothing is random? Can we perhaps even be proud of these positive, only seemingly random events?

So is everything that happens on Earth rather random, or do we have a significant part in it? Do we construct the world for ourselves like Pippi Longstocking ("…I make the world—tiddly-pom, just as I like it…" after Astrid Lindgren)? And if we do construct reality, why should we do so in a negative way? Positive would at least be understandable. Are we searching for old patterns in our lives; if so: why should we do that? Buddhists would see it that way. In their philosophy, we all live so often and so long until we have finally dealt with our old patterns, and only then do we get the well-deserved rest in Nirvana. There, at last, is an end to the eternal rebirths and the repetition of familiar patterns. However, from the standpoint of evolutionary interests, an organism—if it were possible—would rather be reborn than reach Nirvana. Moreover, personal interests could also bring potential conflict with this worldview. Personally, I like my life; so my goal should be to avoid Nirvana. Therefore, I will not work through all my old patterns.

But if everything on Earth were to happen by chance, we would have hardly any responsibility for our actions. Most events would then be unpredictable, and my own actions would have little impact on the course of things. Life would be shaped by events beyond our control. Parallel to random and uncontrollable events in evolution (meteorite impacts, volcanic eruptions, floods, earthquakes, plate tectonics), our personal lives would also be marked by an agony in the face of unpredictable events such as illnesses, sudden deaths, accidents, lucky lottery wins, and (similar) chance encounters. Although no one denies that such events exist, I would like to doubt that only such events shape our lives. Such a dominance of random events is, for me personally, not a satisfying worldview. Of course, this view of things is absolutely subjective.

Or do our lives and evolution proceed in a more deterministic way? Is everything predetermined? That was the idea of the "Laplacean demon," who could predict all future events if only all information about the present were known in detail. This idea was formulated by the French mathematician Pierre-Simon Laplace in 1814 in his preface to the "Essai philosophique sur les probabilités." Today, supercomputers allow us to model and partially predict complex systems such as the weather, climate, ecosystems and their changes, the dynamics of stock markets, economies, human behavior, the course of diseases, etc. Is it only a matter of time before supercomputers can predict our future precisely? But even in determinism, our actions are ultimately meaningless, since everything is already predetermined. In determinism, we bear no responsibility because there is no freedom of action; and ultimately, we cannot incur guilt, since we have no room for maneuver. With determinism, any criminal could excuse themselves. He/she simply could not have acted otherwise.

Natural scientists have quite a bit to contribute to all these questions. Physicists and biologists have been investigating the significance of chance and determinism for many centuries. Classical mechanics states that the apple always falls down from the tree; it never falls up or to the side; so there is a deterministic element in our everyday life that we all know. Most of the time, everything happens as usual, and our perception fits well with events on the meso-level (that is, the level of daily human experience; quantum physics and astronomy are not usually included). In contrast, physics can also show that we do not live in a deterministic world; for many events are unpredictable in detail, e.g., radioactive decay, the behavior of complex systems. The chaos theory of the 1980s showed that there are, in principle, limits to predictability, since tiny local events can unpredictably affect a complex system in a non-deterministic way (Gleick 1987).

What do biologists say about chance and determinism? In evolution, there have certainly been many random events (meteorite impacts, volcanic eruptions, mutations) that have significantly influenced the course of things. However, the relative importance of chance among evolutionary biologists is quite controversial. Stephen Jay Gould, an American evolutionary biologist, considers the role of chance to be essential. In his view, there are no clear lines or directions in evolution. If life were to start again on Earth, the outcome could be entirely different. Some of his books, such as "Wonderful Life" or "Burgess Shale," underpin his philosophy. The interpretation of the Burgess Shale fossils, which are about 560 million years old, plays a central role here. While the first paleontologists who examined this site (Whittington 1975) could not assign these fossils to any known taxon, Collins et al. (1983) concluded that they could all be classified within known animal phyla. Stephen Jay Gould re-examined the fossils and saw in them entirely randomly arising and disappearing taxa, which were neither better nor worse adapted than other lineages (contingency theory of evolution). In this way, he underscored the importance of chance in evolution. Simon Conway Morris sees the course of evolution quite differently. In his view, problems in evolution are solved in always similar ways (convergence theory of evolution). This is of great importance for ecologists and evolutionary biologists. If you enter the search term in Google Scholar (the search engine for scientific publications), you will find about 152,000 papers. In 2005, Conway Morris provided many surprising examples of convergent evolution in his book "Life's Solutions." According to Conway Morris, evolution on Earth would proceed similarly to the present if it were to start over.

The discussion about the Burgess Shale fossils ultimately shows that this is a futile debate. On the one hand, the fossils do indeed look very strange when compared to modern species (especially *Opabinia*, *Marella*, and *Hallucigenia*, to name just three genera); on the other hand, they are still fairly clearly recognizable as arthropods and are therefore not complete mythical creatures to a modern biologist. If one wishes, one can interpret these fossils as having arisen by chance, with some of them also becoming extinct by chance; on the other hand, one can also see them as poorly adapted species that became extinct in the course of evolution (and this was then not a random process). Presumably, there are always phases in evolution of more rapid and slower radiation, which were coupled with the respective biotic and abiotic conditions on Earth. Convergent evolution certainly plays a role in evolution, but that by no means implies that, after a restart of evolution, the same scenario would unfold as in the first run.

The scientific views of the protagonists Stephen Jay Gould and Simon Conway Morris were and often still are closely linked to their personal worldviews. Stephen Gould was an atheist scientist with socialist ideas about justice and a focus on the random course of history and evolution (Gould and Lewontin 1979); Simon Conway Morris is also a biologist, but unlike Gould, he is religious and derives his religiosity at least in part from his research findings. So what do biologists generally say about chance and determinism? Overall, there is no clear picture, and it does not take much to concede a little to both sides. There are evidently recurring convergent events in evolution (i.e., constraints), deterministic elements; problems are solved in similar ways again and again (so does life on other planets ultimately look similar to that on Earth?), and some people draw religious conclusions from this. On the other hand, there are also chance events (Burgess Shale radiation of complex metazoans, mass extinctions, the evolution of us, *Homo sapiens*), which rather give the impression that chance events can play a major role in evolution.

All in all, neither physics nor biology can answer the question of whether the course of events is more deterministic or whether chance influences the course of history and evolution. However, both physics and biology can refute the two extreme positions (everything happens by chance/everything is predetermined); for there is, in any case, an element of chance at many levels in physics and also in biology (some of which we have mentioned above); and at the same time, deterministic factors, classical mechanics, and constraints in evolution (convergent developments) certainly play a major role. Physicists and biologists can say little about the relative importance of chance and determinism, but they can rule out that life, the development of the cosmos, and certainly the fate of humanity have proceeded exclusively by chance or exclusively deterministically. Both factors play a role in the course of the history of the universe and of life.

What do religions and many spiritual thinkers say about chance? Of course, there is no unified perspective here either, but generally, chance presents a problem for all religious thinkers and spiritual people. The role of chance is usually marginalized. Chance must not play a major role, since there is usually a great guide who started the universe and also directs its fate. Many spiritual people give little room to chance, because they often think that an event did not occur by chance, but wants to tell them something or had to happen in some way. In German, there is the term "Schicksal" (fate); in Arabic, one also says Kismet (the predetermined course of events in a person's life) or 'inshallah' ('if God wills'). The flip side of this view, however, is that there is hardly any responsibility and certainly no guilt in such a conceptual framework. We should therefore be grateful for the random events of this universe, for they give us the freedom not to be puppets, but to make our own decisions. Only in a non-deterministic universe is there genuine discourse, real freedom of choice, and the possibility to change and move things. Lawrence of Arabia rebelled early on against the religious/spiritual delusion of denying free will. When, in July 1917, everyone advised him against crossing the part of the Nefud desert considered impassable, over which he wanted to capture Aqaba (now in Jordan) by surprise, Lawrence of Arabia uttered the famous sentence: "Nothing is written." With this, he rejected the religious fatalism of the course of the world. His famous sentence stands as a symbol for human willpower and for overcoming an allegedly unchangeable fate.

So are random events good or bad; is determinism good or bad? I am glad that the universe does not run deterministically, because then I would only be a puppet of higher powers and would have no freedom of choice. Life would seem meaningless to me in a deterministic universe. I am glad that sometimes there are simply coincidences that I cannot influence. They relieve me of the burden of being fully responsible for every event. Sometimes there is simply luck or misfortune in life. So sometimes I can also relax; I will never be able to completely influence all events. Even Lawrence of Arabia would see it that way—or at least I think so. Only in a universe that lies between chance and determinism does our action have meaning.

This brings the arc of this book full circle. I want to analyze, on the basis of the natural sciences, how organisms interact with each other and with their environment, and whether an ethics can be derived from this. I also want to give voice to the side of religion and spirituality. My question is: can a natural ethics and perhaps even politics be derived from the natural sciences? Or is that simply inappropriate biologism? Can we derive anything from nature for our behavior or our morals? Or do we need an ethics, religions, or at least spirituality to prescribe our behavior and morals? Or do we perhaps behave quite reasonably/morally even without higher laws? It is very important to me not to neglect the spiritual perspective, because there are some things in life that elude examination by the natural sciences; perhaps even most important things. Deep down, we are all often very spiritual; for if we were only rational and reasonable, we would almost never act as we often do.

Especially outsiders, that is, non-scientists, often like to draw a sharp line between the logic, reason, and apparent proofs of scientists on the one hand, and the supposedly irrational world of emotions and feelings on the other. But I would like to point out that scientists cannot prove anything and do not deliver truths. All that the natural sciences do is to continually propose new (slightly better) hypotheses about the world, which hold until even better ones are found (loosely following Karl Popper (1934) and Kuhn (1996)). This is precisely the strength of the natural sciences: that they always allow for discussion and that discourse is possible. If there were irrefutable truths that could not be questioned, we would be in the realm of religions (where dogmas must not be questioned) or in the axiomatic realm of mathematics (which, as is well known, is not a natural science). If there are any "truths" at all, they are often found in the emotional domain. If I have a certain perception of a situation, that is definitely my truth. No one can verify this truth, since to this day we are not able to understand the workings of the brain 100%, but no one will (or should) doubt their own feelings. An 'exception' is perhaps Mark Cohn & Blind Boys of Alabama in the song "Walking in Memphis" with the line where they sing "…but do I really feel the way I feel?" This line makes us smile because, after all, we are usually quite clear about our feelings.

In my opinion, the key to understanding the interaction of organisms with their own kind and their environment is also the key to understanding our behavior. The discipline that deals with this is ecology. It studies the behavior of animals and humans and tries to understand this behavior. Therefore, this book will be about the ecology of animals, plants, and humans.

Theodosius Dobzhansky wrote in 1973, "Nothing in biology makes sense except in the light of evolution." I would like to add the following: Ecological interactions shape organisms and are at work in the short term; only in the long term does this then manifest as evolution. Therefore, one could also add to Dobzhansky: "Nothing in evolution makes sense except in the light of ecology." The next chapter therefore begins with the first steps toward understanding ecology; namely, with the significance of abiotic factors, biotic factors, and the importance of the resource factor, which accompany and sometimes end the lives of animals and plants.

References

Alvarez W (1997) T. rex and the crater of doom: Princeton, Princeton University Press

Alvarez LW, Alvarez W, Asaro F, Michel HV (1980) Extraterrestrial cause for the Cretaceous-Tertiary extinction. Science 208:1095–1108

Collins D, Briggs D, Conway Morris S (1983) New Burgess Shale Fossil Sites Reveal Middle Cambrian Faunal Complex. Science 14:163–167

Conway Morris S (2005) Life's solution: Inevitable humans in a lonely universe. Cambridge University Press

Cooper A, Penny D (1997) Mass survival of birds across the Cretaceous- Tertiary boundary: molecular evidence. Science 275:1109–1113

Dobzhansky T (1973) Nothing in biology makes sense except in the light of evolution. Am Biol Teach 35:125–129

Fawcett JA, Maere S, Van de Peer Y (2009) Plants with double genomes might have had a better chance to survive the Cretaceous-Tertiary extinction event. Proc Natl Acad Sci 106:5737–5742

Gleick J (1987) Chaos. Random House

Goderis S, Sato H, Ferrière L, Schmitz B, Burney D, Kaskes P, Vellekoop J, Wittmann A, Schulz T, Chernonozhkin SM, Claeys P, de Graaff SJ, Déhais T, de Winter NJ, Elfman M, Feignon JG, Ishikawa A, Koeberl C, Kristiansson P, Neal CR, Owens JD, Schmieder M, Sinnesael M, Vanhaecke F, Van Malderen SJM, Bralower TJ, Gulick SPS, Kring DA, Lowery CM, Morgan JV, Smit J, Whalen MT (2021) Globally distributed iridium layer preserved within the Chicxulub impact structure. Science Advances 7, eabe3647

Gould SJ, Lewontin RC (1979) The spandrels of San Marco and the Panglossian paradigm: a critique of the adaptationist programme. Proc R Soc Lond B 205:581–598

Halliday TJD, Upchurch P, Goswami A (2016) Eutherians experienced elevated evolutionary rates in the immediate aftermath of the Cretaceous-Palaeogene mass extinction. Proc R Soc B 283:20153026

Hubbell SP (2001) The unified neutral theory of biodiversity and biogeography. Princeton University Press, Princeton, NJ

Kuhn TS (1996) Die Struktur wissenschaftlicher Revolutionen. Suhrkamp

Longrich NR, Tokaryk T, Field DJ (2011) Mass extinction of birds at the Cretaceous-Paleogene (K–Pg) boundary. Proc Natl Acad Sci 108:15253–15257

Pachl P, Uusitalo M, Scheu S, Schaefer I, Maraun M (2021) Repeated convergent evolution of parthenogenesis in Acariformes (Acari). Ecology and Evolution 11:321–337

Popper K (1934) Logik der Forschung. Julius Springer

Raup DM (1992) Der Untergang der Dinosaurier. Rowohlt

Schaefer I, Norton RA, Scheu S, Maraun M (2010) Precambrian mites colonized land and formed parthenogenetic clusters. Mol Phylogenet Evol 57:113–121

Thornton IWB (1996) Krakatau: The destruction and reassembly of an island ecosystem. Harvard University Press

Whittington HB (1975) The enigmatic animal Opabinia regalis, middle Cambrian, Burgess Shale, British Columbia. Phil Trans R Soc Lond B, 2711–2743

Abiotic Factors, Biotic Factors, and Resources

Abiotic factors shape and limit the lives of animals and plants, especially when they are extreme. Low temperatures or extreme heat/drought restrict the distribution of many species. In Antarctica, there is no continuous plant cover (although there are some grasses, mosses, lichens, and a few 'higher' plants), so there are also almost no herbivores (phytophages) there. Nevertheless, some tiny invertebrates such as mites, springtails, nematodes, tardigrades, and rotifers are still found there. How do these animals survive the cold temperatures, and what do they live on when primary production by plants plays almost no role?

Here we encounter an important surprise in ecology. An inexperienced ecologist might think that ecosystems are structured as follows: In every ecosystem, there are plants that produce the biomass in the first place (=primary production). This then allows for their consumers (phytophages) as well as their consumers (predators, parasites); and when the plants die, dead organic material is produced, which is mineralized by decomposers so that nutrients are returned to the plants. While this idea holds true for most ecosystems, it is not always the case. Some systems can (at least temporarily) exist without their own primary production. How is that possible?

Apparently, there are ecosystems, or better, sub-systems—since they lack their own primary production and are thus dependent on other systems—that live off so-called allochthonous resources. This means that the system's resources are supplied from outside in some form (as opposed to autochthonous resources, which originate within the system itself). Typical systems with little or no primary production, and therefore supplied from outside, include rivers (resources come, for example, from surrounding trees whose leaves fall into the water), tidal flats (resources come at least in part in the form of sedimenting algae), glacier forefields (dead organic material is blown in), and indeed Antarctica. The highest densities of invertebrates in Antarctica are found where there are colonies of

M. Maraun, *Everything is Ecology*, https://doi.org/10.1007/978-3-658-50207-2_2

penguins and elephant seals. These animals bring resources from the sea to the land; in terrestrial habitats, this mainly occurs through the production of excreta (feces, urine) (Bokhorst et al. 2019).

The Balance of Nature

I would like to briefly introduce the three factors to which this chapter is dedicated by way of the "schools" that represent each of these factors. The representatives of the 'abiotic school/climate school' are the two Australian biologists Andrewartha and Birch. They laid out the foundations of their ideas in a monumental book (Andrewartha and Birch 1954). Andrewartha studied populations of thrips (Thysanoptera) in Adelaide (Australia) and was able to show that their population fluctuations were primarily shaped by climate and temperature. These populations fluctuated greatly in their densities and gave the impression of being shaped by catastrophic climatic events. The populations of thrips were obviously not dependent on their densities. This regulation by abiotic factors was therefore called density-independent. It means that the mortality rate in the population (or the end of its growth) is independent of the current population size. For example, always 90 (or 80 or 70)% of all individuals die (in the case of drought/dryness/frost), no matter how large the population is at that moment. This did not correspond to the typical image of the "balance of nature" that was widespread at the time.

This image was rather represented by Nicholson, also a biologist living in Australia. He studied the population fluctuations of blowfly (Calliphoridae) larvae parasitizing sheep, from which he derived models for parasite-host interactions. In 1933, he published the paper "The Balance of Animal Populations" in the Journal of Animal Ecology. With his research findings, he established the idea that nature is, in a certain sense, in equilibrium (Nicholson 1933; Nicholson and Bailey 1935). The larvae of his blowflies penetrate the skin of the sheep and feed on their internal organs, so that the sheep die sooner or later. The adult blowflies then need a new host to lay eggs. That Nicholson found a rather density-dependent population regulation in parasitic blowfly larvae is not surprising. Animals that live off a locally limited prey (sheep) perceive the decline of this resource with increasing intensity.

In his experiments with blowflies and their larvae, he provided the animals with either an excess of food or only very small amounts. He was able to show that in experiments with very little food (for the larvae only!), the density of the blowflies fluctuated around a certain value (carrying capacity), because sometimes too few or too many eggs were laid. This depended on the number of adults in the previous generation. When he provided food ad libitum (without limitation), a massive number of eggs were laid; the latter, however, is an extremely rare or only temporary case in nature. In another experiment (realistic for the wild), he provided both adults and larvae with limited food; in this case, the populations slowly grew toward their capacity limit. In this way, he demonstrated the density-dependent regulation of a population. Density-dependent regulation of populations concerns not only parasite-host relationships, but also predator-prey systems. As prey density increases, the predation pressure from predators rises, so that (slowly) a density-dependent regulation of the population occurs. In density-dependent regulation of populations, the population increasingly perceives an emerging problem as density increases; and as a result, it grows more slowly or eventually not at all. This is the crucial point in density-dependent population regulation, in contrast to density-independent regulation, where catastrophe (volcanic eruption, frost) occurs without warning. Density-dependently regulated populations can therefore react to a problem to a certain extent. They can emigrate, they can try to develop protective mechanisms, or they can hide.

In most ecology textbooks, only these two ideas about the regulation of populations are mentioned, and density-dependent and density-independent regulation are indeed the two most important possibilities. But there is another factor that also leads to density-dependent regulation of populations, namely the factor "resource." As perhaps the most important representative, one could cite another Australian, T.C.R. White, who argues that resource (he focuses especially on nitrogen as a limiting factor) is the most important factor for almost all organisms. He presented his idea in publications and in the book "The Inadequate Environment" (White 1993). Even in the case of resource scarcity, populations are regulated density-dependently (the more resource is consumed, the more slowly the population grows). Therefore, the representatives of the 'Resource School' can also be assigned to those ecologists who view nature as a system that is usually in equilibrium. Typical examples of the great importance and limitation of resources are phytophagous ungulates such as African Cape buffalo. They feed on plant material (grasses) in the open savannas of East Africa, which enables their nutrition and growth up to a certain point; but if the nitrogen content of the grasses is too low, they can no longer use them as a resource. The same restriction applies to invertebrate phytophages. This shows that resources, or resource accessibility and quality, are a crucial key to understanding the regulation of populations. I would like to emphasize at this point that, in what follows, I favor this resource theory, without ignoring exceptions and without being dogmatic.

The density-independent school of Andrewartha and Birch (1954) clearly distinguishes itself from the other two density-dependent schools; but the separation of the two density-dependent theories is not so simple. The crucial difference is that, according to the resource school, a population is regulated density-dependently because it increasingly depletes a resource. In Nicholson's school, on the other hand, a population can also be regulated density-dependently (e.g., by a predator; the more prey there is, the more the predator eats); but the predator does not become rarer when its prey becomes less abundant (e.g., because as a generalist it switches to another type of prey; only specialized predators suffer when their prey becomes rarer). In the resource theory, therefore, consumer and resource are closely coupled; in the biotic school, the regulator and its prey (parasite-host/predator-prey) are not necessarily coupled. Especially not if the consumer is a generalist.

The respective factors that regulate populations density-dependently or density-independently are also not always easy to distinguish. An important factor for the growth of plants is sunlight: is this now an abiotic or a biotic factor, or a resource? Here, we define a resource as a factor that becomes depleted when it is used. Such factors are usually also competed for. Light is essential for plants, and they do compete for it (the fact that light comes from above is the reason for the height of trees; they have evolved to be so tall in order to outcompete their rivals for it); but sunlight does not become weaker when plants use it. Thus, light is a mixture of an abiotic factor (it is not changed by use), but in the long term in evolution it certainly has a facet of a resource (since plants compete for it, and for plants growing beneath trees, light can be a limiting factor).

Many factors can be resources that one would not expect. Of course, all things that can be eaten (and that do not regenerate as quickly as they are consumed) are resources, but space can also be a resource for sessile organisms (i.e., plants and sessile marine animals), or mating partners. Sometimes all these factors are not clearly separated. What is an excretory product for one (ammonium, urea, urine, feces) is a resource for another (plant, coprophagous animal). There is certainly some truth in all theories; in fact, it depends on the organism, its habitat, and its interaction with other organisms.

These three schools form the basis for working with the most important factors in ecology: abiotic factors such as heat, cold, seasonality, storms, earthquakes, volcanic eruptions; biotic factors such as intraspecific competition, territoriality, diseases, predation; and the resource factor. Biotic interactions and resources naturally sometimes overlap and—as we have seen—are often difficult to separate. After this introduction, we will now turn to a selection of specific abiotic environmental factors and their effects on the animal and plant world.

Temperature

Temperature is one of the most important factors influencing the activity of both animals and plants. Global temperature differences determine the distribution of animal species as well as their overwintering strategies. While in northern and southern latitudes most invertebrates overwinter as eggs, this issue is irrelevant in the tropics. On the other hand, in the tropics, temperatures can fluctuate greatly over the course of a day; organisms must adapt to this. Organisms can only be physiologically active within a certain temperature range; outside of this range, they enter cold torpor or die, or their proteins degenerate at excessively high temperatures. Only a few microorganisms can survive permanently at temperatures above 100 degrees. When considering the factor of temperature, it is useful to return to the initial example and ask: how do animals survive the cold temperatures of Antarctica? The emperor penguins living there have dense plumage that is coated with a secretion from the uropygial gland, making it completely waterproof. The fat layer of emperor penguins also contributes to thermal insulation. In addition, these animals form so-called huddling groups (especially after sunset) to protect themselves from the cold.

But how do tiny invertebrates survive temperatures of –70 degrees and below? Invertebrates cannot protect themselves from the cold with feathers, fur, or thick layers of fat. To a certain extent, they are frost-tolerant by producing and storing glycerol (alcohol), glycogen (polysaccharide), sorbitol (sugar alcohol), or antifreeze proteins in their bodies. In this way, they too can survive temperatures as low as −30/−40 degrees (Willmer et al. 2005).

Sometimes abiotic factors also interact with the availability of resources. During cold periods, resource availability is often reduced (especially for terrestrial animals). Organisms have developed different strategies to survive such phases. Small warm-blooded (endothermic, homeothermic) animals such as marmots, badgers, bats, opossums, and hummingbirds enter a state known as torpor, which can also be described as hunger or cold sleep. Torpor is usually triggered by food scarcity. These animals live off fat reserves they have accumulated and are more common in productive, boreal systems than in the unproductive polar regions such as the tundra, where they could hardly accumulate fat reserves. Torpor is a strong effect. All physiological processes are reduced, the animal barely responds to stimuli; however, torpor usually lasts only a short time (hours to weeks) and endothermy and thermoregulation are not abandoned (the animals' temperature remains constant at about 5–15 °C). Large animals such as skunks, hedgehogs, beavers, and bears, on the other hand, enter hibernation, which can be seen as a mild form of torpor. Their physiology is only slightly reduced and they become active when they are hungry. Hibernation can last a very long time (up to 7 months in the edible dormouse), interrupted by a few phases of foraging. A third way to survive periods of cold and hunger is cold torpor or winter torpor. This is mainly used by animals that are capable of it, usually cold-blooded animals such as invertebrates (snails, insects), amphibians, reptiles, and fish. In this phase, the animals are usually immobile, have their eyes open, and do not eat. An increased glucose concentration can prevent lethal freezing to a certain extent. This phase can last a long time (pond newts (3–4 months), slow worms and adders (4–5 months), tree frogs and sand lizards (5–6 months)).

In connection with extremely cold and hot habitats, Bergmann's rule is often mentioned, which many of us learned in school. It states that animals become larger toward the poles because their skin surface area relative to body volume becomes smaller. This is true for the emperor penguin, the largest living penguin, but not for the tiny invertebrates in Antarctica, which are extremely small and protect themselves from freezing with various chemical substances (see above). Thus, the observation is rather that animals become either larger or smaller toward the poles (at least in Antarctica). Only animals of intermediate size are absent; they apparently have the greatest problems with the cold.

In connection with Bergmann's rule, Allen's rule is often mentioned, which states that animals have smaller and shorter extremities toward the poles to lose less heat. In contrast, extremities are relatively large in desert regions to dissipate heat. The classic example is the arctic fox with tiny extremities, the red fox of temperate latitudes, and the South African fennec, a desert-dwelling fox with large extremities. Unlike Bergmann's rule, this rule also seems to apply to invertebrates. In ground-dwelling mites and collembolans, antennae and legs are shorter in cold regions. Apterous or brachypterous (wingless or short-winged) flies and mosquitoes are also more common at the poles to avoid heat loss through the wings (Willmer et al. 2005). There are probably other reasons for the loss of wings in such island habitats (true islands or habitat islands). Dispersal in such habitats is almost always fatal and therefore not sensible; thus, wings can also be eliminated, as they are more of a hindrance and consume resources for their development.

An important ecological rule related to temperature for organisms is the so-called Q10 rule (reaction rate–temperature rule; or Van't Hoff's rule). It states that the reaction rate of organisms doubles when the temperature increases by 10 degrees. Of course, this rule does not apply to the entire range of existence of organisms; it is mainly relevant at low temperatures; at high temperatures, the effect is smaller. Therefore, the current increase in global average temperature is of particular importance in cold habitats.

Light

The process that enables almost all life on this Earth (with the exception of some systems with sulfur-autotrophic bacteria; e.g., some caves and the "deep sea smokers") is light-dependent photosynthesis. The biomass produced by plants is called primary production (in contrast to the secondary production of animals, which feed on plants or on herbivores). Since all heterotrophic organisms depend directly or indirectly on organic compounds to grow, light is one of the most important factors for life on Earth. If photosynthesis temporarily fails or is reduced (as after meteorite impacts or volcanic eruptions, when local or global darkening occurs), this has catastrophic effects on the organisms of this Earth.

Photosynthesis is a biochemical process and therefore only slightly dependent on temperature. This is of great significance for the ecosystems of this Earth. It means that even in relatively cold habitats, a great deal of photosynthesis still takes place. The process that can be seen as the counterpart to photosynthesis is the microbial decomposition of dead organic material. This process is what makes life on Earth possible in the first place. Without decomposition processes, dead organic material would accumulate, and organisms would have difficulty accessing nutrients that are locked in this dead organic material (we will return later to processes that, for thousands of years, prevented the steady decomposition of dead organic material; during this time, coal, oil, and natural gas were formed on this planet…). At present, however, we are not accumulating dead organic material in our ecosystems, and thus not accumulating carbon ("we are not making coal"; Hairston et al. 1960), which is the basis for the topic of global change (next chapter), and therefore, at the moment, the ecosystems of this Earth can be regarded as being in equilibrium between primary production and decomposition (more or less, at least).

Microbial decomposition processes, however, are more strongly temperature-dependent than photosynthesis. This means that in the tropics, microbial decomposition processes proceed more rapidly, so that hardly any dead organic material accumulates there. In the cold biomes of this Earth (tundra, taiga), however, strong photosynthesis and low decomposition lead to an accumulation of dead organic material. This material would be released in the event of a temperature increase on Earth, since in such a case the microbial processes would proceed more quickly; this is currently being discussed as one of the major problems of global change/global warming.

The largest part of the light, however, is not used by plants at all; only about 1% of the incoming energy is used by plants. Light is therefore usually not a limiting factor for plant growth and thus for primary production. However, light and its utilization have certainly been an important evolutionary driving force in the development of plants. The competition among plants for light can be seen in the structuring of a forest into different layers; it also led to specialized sun and shade plants. Light (or its absence) also determines the periods of activity of diurnal and nocturnal organisms. Much more important (in terms of limitation) for plant growth, however, are probably CO_2 and nutrients (e.g., N, P, Ca, Mg).

Ion Concentrations (Osmolarity)

Osmoregulation is the ability of organisms to regulate their osmotic pressure. This pressure arises because (especially aquatic) organisms have a higher or lower ion concentration than their environment. Animals that are iso-osmotic have the same ion concentration as their surroundings. Freshwater fish tend to absorb ions (they are so-called hyper-osmotic regulators); saltwater fish tend to excrete ions (thus, they are hypo-osmotic regulators). Freshwater is typically characterized by a low ion content, whereas saltwater is rich in ions (salts). To maintain a certain osmolarity (ion concentration) in their bodies, freshwater fish must absorb ions (e.g., Na, K, Cl), while marine fish (mostly bony fish) excrete ions through their skin (e.g., Na, K, Cl) and via their urine (Mg, S).

But why do fish do this? Cartilaginous fish (sharks, rays, and chimaeras) are iso-osmotic. That is why we rarely eat sharks, as their flesh is very salty; except sometimes "Schillerlocke," which is the smoked belly flap of the spiny dogfish. However, Schillerlocke is very salty and, like many other top consumers (predators), contains a lot of mercury, and because the spiny dogfish is endangered, Schillerlocke is rarely available on the market. Other sharks taste of ammonia or urea (because they regulate their ability to ascend and descend via their urea content) and are therefore usually inedible. If one wants to eat shark (as Icelanders do with the Greenland shark), it must be stored openly for months so that the urea can first be converted into ammonia, and then the ammonia can evaporate.

The fact that the marine fish we eat (e.g., cod, herring) are by no means as salty as cartilaginous fish suggests that this group has an evolutionary background in freshwater. They consist largely of water with low ion concentrations (essentially drinking water; pirates of the seas could have used them as a source of drinking water at any time if they ran out). This hypothesis is consistent with the phylogeny of the ray-finned fishes (a subgroup of bony fish). They most likely migrated from freshwater into marine environments. Today, some of these taxa still live in freshwater, for example, bowfins, bichirs, gars, and sturgeons (though only a few species remain; today <30), while the subsequent radiation of marine fish (Teleostei) was very intense and led to about 30,000 species, of which today over 60% live in seawater and only 40% in freshwater.

Osmoregulation itself is a challenge for many fish in both freshwater and saltwater. It is presumably even more difficult to adapt to the fluctuating salinity in brackish water or to transition from living in freshwater to saltwater. Overall, far fewer fish species live in brackish water than in saltwater; only a few species migrate back and forth between freshwater and saltwater. However, some species (about 160 out of 30,000 fish species) have managed to cope with this problem. These are the so-called anadromous (Greek ana = up; dromos = to run) and catadromous (gr. kata = down) species. The key point is the location of spawning. Anadromous fish such as sturgeons, lampreys, and Pacific salmon spend most of their lives in the ocean and migrate (upstream) into freshwater to spawn. Many animals, such as bears and humans, catch these fish shortly before spawning (on their way upstream), because they are then large and represent a good resource. Catadromous species such as eels, mullets, some perches, and sculpins spend most of their lives in freshwater and migrate to the sea to spawn. In contrast to anadromous species, we catch and eat adult eels on their way to the sea, because they are then large and nutritious. Eels can also be caught and eaten as small young eels, so-called glass eels, on their way into freshwater. However, due to their small size, they are not a significant food source. But why do these species migrate at all? And why do they migrate in different directions? To understand these migrations, one must look at where these species are mostly found and how resource availability is for these fish. Anadromy is a phenomenon of northern latitudes, while catadromy is more a phenomenon of the tropics.

The most important point is this: Where there are many resources (=preferred habitat), there are unfortunately also many predators; and where many animals reproduce, there are even more resources—namely, their eggs. Certain salmon have tasty and nutritious eggs (think of images of bears catching salmon), which they prefer to deposit in a habitat where there are not so many predators. Such a habitat is the nutrient-poor rivers of the northern latitudes. Therefore, salmon migrate from the nutrient-rich northern marine areas into the nutrient-poor, predator-poor environments of the rivers in northern regions.

Eels, on the other hand, migrate to spawn in the Sargasso Sea, a vast nutrient-poor gyre in the northwest Atlantic, bounded by the Gulf Stream to the north, the North Atlantic Current to the east, and the Canary/North Equatorial Current to the south. There, they spawn at depths of up to 2000 m, in a presumably almost predator-free environment, and die. The small glass eels migrate over the course of about three years back to 'their' rivers in Europe, feed there, and become the large silver eels, which then migrate again to spawn towards the Atlantic.

Why are these migrations of eels and salmon so opposite? To answer this, one must know that eels are not typical representatives of catadromous species. Fundamentally, however, they do have a similar strategy to the anadromous species (i.e., salmon). They seek resource-rich areas for feeding and resource-poor areas for spawning (to avoid predation). Catadromous species thus "normally" feed in the nutrient-rich areas of tropical rivers. The typical whitewater rivers of the tropics, such as the Amazon, are richer in nutrients than rivers of temperate latitudes, because they originate in recently geologically active regions and therefore carry a large amount of eroded, turbid material. The suspended matter is often rich in iron and aluminum; they also contain many electrolytes and nutrients. Their spawning grounds are located in the nutrient-poorer marine areas of the tropical and subtropical zones. Conversely, anadromous species (i.e., salmon) feed in the polar, nutrient-rich marine areas (upwelling zones) and spawn in the relatively nutrient-poor rivers of temperate latitudes. In terms of resource availability and predation avoidance, the different migrations of anadromous and catadromous species are easy to understand. Is this just 'story telling'? No, there is also solid evidence for this: So-called "land-locked sockeye" salmon, which do not migrate but live in lakes in North America, New Zealand, and Europe, have only about a quarter (!) of the weight of the migratory species, which points to the great importance of migration for the nutrition of these animals. Land-locked sockeye salmon have either been introduced into lakes by humans for fishing, or have been cut off from the sea by geographical events (e.g., lowering of sea level at the end of the last ice age), or can no longer migrate due to dam construction, or have simply ceased their migrations for no apparent reason.

European, American, and Japanese eels are an exception to the above pattern. Eels are actually a rather tropical group; their migrations should therefore actually end in nutrient-rich tropical rivers, where they should feed. Apparently, however, European rivers are also sufficiently nutrient-rich for eels to enable growth there.

Acidity (pH Value)

The pH value is also an important factor for ecosystems. It is a measure of the acidity of a soil (it also exists in oceans, but that is not the topic here). Relatively high pH values are typical for calcareous soils, which often originate from marine deposits, for example with shell limestone as the parent rock. Such soils host a large proportion of fauna that depend on calcium for their body structure or metabolism, such as millipedes, woodlice, earthworms, and snails. The lower the pH value, the more H^+ ions are present in the soil and bind to the soil's ion exchangers, where they displace Al and Fe ions. These Fe and Al ions are toxic to animals and fine roots and can lead to the phenomenon of forest dieback. I was socialized in ecology with the associated topic of acid rain. Back then (1980s/1990s), there was no other ecological topic in the media. Der Spiegel headlined "the forest is dying," there was alarmism and catastrophism; Nostradamus was quoted, and the rest of the world spoke somewhat pityingly of "The German Waldsterben," even though other countries were also affected (Austria, Czech Republic, etc.).

In the end, the forest did not die; technical countermeasures were taken (desulfurization plants, unleaded gasoline), and the forest recovered. The press had to face accusations of catastrophism (many pictures of dying forests were taken in areas where the forest had always naturally struggled to grow; e.g., acidic soils at high altitudes; Ackerbruchberg in the Harz, etc.), and the apocalypticists had to wait many more years until the forest died again. Overall, this was a success story for nature conservation, and it was obvious that the damage caused by humans could also be repaired by humans. Forest dieback due to acid rain is no longer a major issue today. Overall, the forest area in Germany has remained relatively constant over the last 200 years and is about 30%, with a slightly increasing trend (Poschlod 2015).

But nowadays the forest is dying again; this time not due to acid rain, but due to drought (global warming). The reports in the press are again similarly alarmist as during the forest dieback. In this case, however, the problem will not be as easy to solve as with acid rain, since trees can only adapt to severe drought to a very limited extent in a short period of time. Moreover, the problem must not only be solved locally (as was at least partially possible with acid rain; that was mainly a problem of highly industrialized countries), but is a global challenge (we will return to this in the chapter "tragedy of the commons").

Oxygen

Another significant abiotic environmental factor is oxygen. About 20% of the air today consists of oxygen. Oxygen probably formed on Earth about 3.5 billion years ago through the activity of cyanobacteria (either free-living cyanobacteria or those that formed so-called stromatolites). Its concentration rose significantly from virtually 0% about 2.4 billion years ago, culminating in the so-called oxygen catastrophe. Oxygen is lethal to anaerobic microorganisms and at that time led to a mass extinction of anaerobes. Today, oxygen is of essential importance for the respiration of almost all animals; and life on this Earth would certainly look completely different without the massive release of oxygen by early microorganisms. It is questionable whether complex organisms such as metazoans could have formed without oxygen. Metazoans take up oxygen either through pulmonary respiration, tracheal respiration, gill respiration, or cutaneous respiration. A few invertebrates also breathe using plastron respiration or a physical gill. The common backswimmer (*Notonecta glauca*), for example, breathes using a physical gill. In this process, air is taken underwater and the oxygen from this bubble is consumed. However, there is pressure on this bubble underwater, causing it to become increasingly smaller, so it must be renewed from time to time, for which the backswimmer briefly surfaces. In plastron respiration, a few (very few) species (e.g., water bug, riffle beetle) prevent pressure from being exerted on the air bubble. To do this, they form a stable outer layer that prevents the bubble from shrinking. The oxygen used by the animals diffuses from the water into this bubble, so the animals never have to surface. However, this method requires very oxygen-rich waters, so there are not many taxa that use plastron respiration. In oxygen-poor (anoxic) areas (e.g., bottoms of bodies of water; Wadden Sea from just a few centimeters depth), there are almost no organisms that can live there except anaerobic bacteria. Only a few metazoans possess symbiotic, anaerobic bacteria that enable them to survive in anoxic areas.

Carbon Dioxide

Carbon dioxide (CO_2) is taken up from the atmosphere by plants and a few photosynthetic bacteria (e.g., cyanobacteria) and converted into carbohydrates with the help of photosynthesis (plus light and water). This process provides almost all of the Earth's biomass (primary production). CO_2 is now released into the air and thus into the atmosphere in large quantities when fossil fuels (coal, gas, oil, oil shale) are burned, where it acts as a greenhouse gas. This automatically makes ecology political; for one of the most important questions in ecology is whether the additional CO_2 we release from fossil fuels can be taken up by plants or not (this would be the case if CO_2 were a limiting element for plants). If CO_2 were not a limiting element (but, e.g., nutrients were more important for plant growth), we could not hope that the CO_2 we release would be taken up by forests. Questions like these show the great importance that ecological research can have for political decision-makers (whether they then base their decisions on this research or ignore it is no longer in the hands of natural scientists).

Ultimately, laboratory and field experiments show that young, growing plants that grow under good conditions without nutrient limitation are able to take up additional carbon, while most old forests (which make up the bulk of the world's forests) are not able to take up excess CO_2, as their further growth is more limited by nutrients and by competition for these nutrients with microorganisms and other trees (e.g., in the "Swiss Canopy Crane Project"). In fact, we cannot hope that plants (especially trees) will take up our excess CO_2. So where should the CO_2 go? How can we bind all the CO_2 that we release from fossil fuels (if we want to)? And once again, ecology becomes a political issue (just as acid rain once was). By talking about CO_2, we are, of course, addressing one of the most current topics in ecology, global change, which we will discuss in more detail in the next chapter.

Abiotics, Resource, and Reproduction

For the organisms of this Earth, all these abiotic factors are of central importance. Depending on the manifestation of these factors, living beings adapt in a wide variety of ways, especially to extreme forms of these abiotic factors. Thus, across the globe, one finds diverse behaviors that are related to the abiotic environment to which organisms are exposed.

For example, one can look at the lifestyle and reproduction of certain Antarctic species, e.g., elephant seals. The males (bulls) compete for the ten to twenty females (cows) of a harem. This leads to dangerous fights in which the bulls sometimes seriously injure each other. These fights can last for weeks, as the defeated bulls repeatedly attempt to mate with the cows. The rearing of young is also rather atypical in emperor penguins. The female lays an egg and then goes off in search of food. The male places the egg on its feet, covers it with its abdominal fold, and carries it around in this way. So, do animals in extreme climates exhibit unusual reproductive patterns?

In desert habitats, a different pattern emerges. More species than usual are gregarious or social/eusocial, i.e., they live closely together in groups and cooperate with conspecifics to varying degrees. These include termites, social isopods, social spiders, darkling beetles (Tenebrionidae), and in the South African desert, naked mole-rats, which are the only known eusocial vertebrates. So, are only animals that have developed unusual cooperation patterns able to exploit the limited and heterogeneously distributed resources of these habitats shaped by extreme abiotic conditions? In extreme habitats (arid deserts or cold deserts), terrestrial animals are thus more likely to be social and cooperative (due to limited resource availability), whereas marine animals in seemingly extreme habitats (cold marine waters) are not, since their resources are available in large quantities. We will return to this when we look at the migrations of marine animals (e.g., whales) to the nutrient-rich, so-called 'upwelling zones' of the cold polar regions, where they find a large amount of prey (resource).

Another reproductive pattern in extreme habitats (deserts, Arctic, Antarctic) is parthenogenesis. In such habitats, a disproportionately high number of asexually reproducing animals occur, both among vertebrates (Kearney 2003) and invertebrates. It is possible that in habitats with harsh abiotic conditions, parthenogens generally dominate (Scheu and Drossel 2007; Maraun et al. 2019), since there is never an overexploitation of available resources and sex does not pay off (geographical parthenogenesis).

References

Andrewartha HG, Birch LC (1954) The distribution and abundance of animals. University of Chicago Press

Bokhorst S, Convey P, Aerts R (2019) Nitrogen inputs by marine vertebrates drive abundance and richness in Antarctic terrestrial ecosystems. Curr Biol 29:1721–1727

Hairston NG, Smith FE, Slobodkin LB (1960) Community structure, population control, and competition. Am Nat 94:421–425

Kearney MR (2003) Why is sex so unpopular in the Australian desert? Trends Ecol Evol 18:605–607

Maraun M, Caruso T, Hens, J, Lehmitz R, Mumladze L, Murvanidze M, Nae I, Schulz J, Seniczak A, Scheu S (2019) Parthenogenetic vs. sexual reproduction in oribatid mite communities. Ecol Evol 9:7324–7332.

Nicholson AJ (1933) The balance of animal populations. J Anim Ecol 2:132–178

Nicholson AJ, Bailey VA (1935) The balance of animal populations. Part I. Proc Zool Soc Lond 3:551–598

Poschlod P (2015) Geschichte der Kulturlandschaft. Ulmer, Stuttgart

Scheu S, Drossel B (2007). Sexual reproduction prevails in a world of structured resources in short supply. Proc Roy Soc Lond B, Biol Sci 274:1225–1231

White TCR (1993) The inadequate environment: Nitrogen and the abundance of animals. Springer, New York

Willmer P, Stone G, Johnston I (2005) Environmental physiology of animals. Blackwell Science

Biologism: Is Nature Good or Evil?

Some spider wasps (Pompilidae) have an unusual way of providing for their offspring. While adult spider wasps feed on pollen, the larvae eat spiders. At first, this is not so surprising, since many hymenopterans—including spider wasps—provision their nests with spiders or insects for their young. To ensure this supply, however, the spider wasps must first catch these spiders; to do so, they actively run and jump around on the ground. When spider wasps find a spider, often a wolf spider (Lycosidae), they position themselves over the spider with their long legs and sting it in the central ganglion to paralyze it. The living but paralyzed spiders are then carried into the wasps' nest and are eaten alive by the larvae over a long period. Other spider wasp larvae live as ectoparasites (external parasites) on free-roaming spiders that have been briefly paralyzed by the adult female wasps for egg-laying. These larvae also consume the spiders alive. They kill the spiders as late as possible to ensure their food remains fresh for as long as possible. From a human perspective, this way of feeding offspring can certainly only be described as brutal.

But isn't nature in general incredibly brutal? We all think of wildlife documentaries in which a cheetah chases a gazelle, or a crocodile performs a death roll in shallow water to drown a wildebeest; we think of the raptors being fed in 'Jurassic Park' or a *Tyrannosaurus rex* attacking humans. We often imagine nature as rather bloodthirsty and merciless; "nature red in tooth and claw," as the Anglo-American would say; everywhere blood, teeth, and claws. However, our empathy for pursued and hunted vertebrates is usually higher than for invertebrate victims—especially when the victims are spiders. Again and again, predator-prey systems or parasitoid-host relationships are seen as exemplary and thus as standard examples of the state of nature. From this, it is then concluded (or secretly thought): that's just how nature is, and that's somehow how we are too; nothing

M. Maraun, *Everything is Ecology*, https://doi.org/10.1007/978-3-658-50207-2_3

can be done about it. Who just caught themselves nodding? And who feels repulsed by the whole story?

With such a brutal view of nature, comparably brutal human actions are (or were) justified, because that's just how nature is. This kind of fallacy (inferring an ought from some supposed state of nature) is also called Social Darwinism. In general, it is a widespread mistaken intuition to infer desirable norms from any given state of affairs that matches the model we observe in nature. By this logic, any even the most bizarre human behavior could be justified, since somewhere in nature there is surely a precedent for it.

However, the question remains whether the entire brutal perspective presented above does not in fact portray nature in a completely one-sided and false way. Isn't nature, in fact, mostly harmonious, cooperative, and symbiotic? Let's look at an example. The tropical rainforests of this earth are, in terms of nutrients, very poor systems. The forests there grow on very old soils, which, simply due to their age, are very nutrient-poor. Nevertheless, we all know films and images of the rainforest where a great diversity of plants and animals can be sustained. Biologists call this phenomenon the "Paradox of Enrichment." In nutrient-rich systems, a few highly competitive species can dominate, whereas in nutrient-poor systems like the tropics, many different species can coexist because no single species is able to outcompete the others. This explains the high diversity of tropical rainforests; but still, some may wonder where the mighty tropical trees get their resources for growth. Carbon for building lignin and cellulose is not a problem, since the trees have enough light and water for photosynthesis; but nitrogen and especially phosphorus are elements that are in short supply in tropical ecosystems. Phosphorus, in particular, is present only in small amounts and is usually hardly available to plants, because in acidic systems it forms complexes with aluminum and iron ions, and in basic systems it complexes with calcium and magnesium. Unlike nitrogen, phosphorus cannot be obtained from the air by microorganisms, because it is not present there. Phosphorus is generally a key element in terrestrial soils, because its total global quantity is much lower than that of nitrogen and is therefore in short supply for many plant species. So where do the plants of the tropics (especially the trees) get their nutrients—especially phosphorus?

To access the few available nutrients in tropical soils, many plants enter into very close symbiotic relationships with fungi; these fungi are called mycorrhizal fungi. In contrast to the loose associations of ectomycorrhizae, which grow only on the outside of plant roots in temperate regions, the endomycorrhizae of the tropics live in a very intimate connection with their plants.

Endomycorrhizal fungi cannot be assigned taxonomically to any of the better-known fungal taxa. They are their own taxon, the Glomeromycota. About 80% of all vascular plants worldwide possess endomycorrhizae, which indicates how important these mycorrhizae are for plants. Endomycorrhizae dominate in the tropics, in grasslands, in steppes, in savannas, and in deserts. Endomycorrhizae are sometimes also called VA mycorrhizae (=VAM); the V stands for vesicular (bubble-forming) and the A for arbuscular (treelet-forming), indicating that these fungi penetrate the cells of the root and form small vesicles and tree-like structures there. In these areas, water and nutrient exchange occurs between the plant and the VA mycorrhiza.

The symbiosis with the plants is very close. The plants depend on the VAM. The fungi increase the root surface area, making water uptake easier; they supply the plants with rare phosphorus (as phosphate) and often also ammonium, as well as zinc and other micronutrients. The fungal hyphae are thinner than the plant's fine roots and can thus penetrate soil pores that would otherwise be inaccessible to the plants. The plant, in turn, provides carbon for the fungus; and over millions of years, the fungus has coevolved with the plant to such an extent that it can no longer absorb carbon from the environment on its own. For both partners, the symbiosis is—except for a few exceptions—obligatory. Without the plant, the fungus cannot survive, and the plant can hardly survive in the nutrient-poor systems of the tropics or grasslands without the VA mycorrhizae. Mycorrhizal symbioses are very ancient. The first land plants already had VAMs. It is possible that endomycorrhizae made it possible for plants to colonize land in nutrient-poor areas in the first place.

VA mycorrhizae are not specialists. This means that a fungus can penetrate root cells of several different plant species. Thus, a fungus can form an underground network of hyphae in which different types of plants are interconnected. Overall, this is a harmonious and close relationship characterized by mutual dependence. Without their respective partners, they perish. Thus, the mutualistic cooperation of land plants and VAMs is one of the most beautiful examples of perfect harmony in nature. Isn't nature ultimately incredibly cooperative, symbiotic, and harmonious; and do we simply not perceive this facet as easily as we do predator-prey systems? The slow growth of a plant together with its root mycorrhiza is simply not as telegenic as lions and antelopes.

When we look at cooperation and symbioses, we see how well living beings can work together. Insects pollinate flowers and are rewarded with pollen; a mutually beneficial cooperation. Lichens are an almost perfect symbiosis of algae and fungi; leafcutter ants have been cultivating fungi for millions of years to the mutual benefit of both; a form of agriculture that we humans also invented 10,000 years ago for the benefit of people and for the benefit of many plants and animals that we have spread worldwide through agriculture. Corals possess so-called zooxanthellae as close symbionts, which are often dinoflagellates that perform photosynthesis and thus provide the coral with organic compounds such as sugars and starches, while in return they receive protection and CO_2. Cleaner fish remove parasites and dead skin from the mouths and scales of other fish at cleaning stations, to the advantage of both participants; one receives food, the other is cleaned. Very close are also the symbioses of nitrogen-fixing root nodule bacteria and legumes (Fabaceae), such as peas, beans, lentils, peanuts, and clover. Before the development of the Haber–Bosch process for industrial nitrogen fixation at the beginning of the 20th century, this symbiosis was almost the only way for plants to get the unreactive atmospheric nitrogen into the soil. In the end, both the plants and the nodule bacteria benefit. In general, these symbioses are widespread in systems where nutrients are scarce. Symbioses become even closer when one partner gives up part of its abilities (which it has stored in the form of DNA), thereby maximizing cooperation with another organism. This is the case with our mitochondria, which we learned about in school as the powerhouses of the cell. These mitochondria were originally bacteria that have transferred most of their genes to the cell nucleus. They rely entirely on their host for nutrition; on the other hand, almost no organism can survive without mitochondria. Cooperation can hardly be closer, more dependent, or more trusting than this.

All these examples can give the impression that the entire Earth is a single cooperative superorganism that cycles nutrients for the benefit of all and ensures every organism's livelihood. This idea is called the Gaia hypothesis (Gaia = great mother in Greek). According to this, the entire Earth is to be regarded as a single living being that produces a habitable biosphere for all its members and thus ensures the survival of all organisms. Under this hypothesis, ultimately all organisms would work together for the good of all. This idea caused quite a stir in the 1970s. It fit well with the era of the hippie movement and stood in stark contrast to the "nature red in tooth and claw" worldview. Ultimately, the Gaia idea is almost a religious worldview, for example, Goethe's pantheism, in which God is seen in all things.

Unfortunately, all these symbioses are not as harmonious and stable as one might think. Mycorrhizae and their trees can be separated. When resources are plentiful, the tree sheds its ectomycorrhizae (this is not possible with endomycorrhizae; their symbiosis with trees is too close).

Many mutualistic systems are open to cheating and betrayal. Cleaner fish do not always remove only parasites, but sometimes also bite off small pieces from the fish they are supposed to clean. Ants milk aphids and protect them from enemies; but if the ants are short on resources, they eat some of the aphids. The yucca moth pollinates the yucca plant and lays some eggs on the flower. However, if the yucca plant is infested with too many moth larvae, it will shed its flower entirely. Orchids deceive pollinators by simulating a mating partner, so that only one participant benefits from the pollination. In this case, there is no pollen for the insects.

If these systems are so open to cheating, one might ask whether symbioses are not rather intended to take advantage of and cheat others (those who do not form symbioses). Isn't the idea behind every cooperation the attempt to outcompete other organisms? Isn't every cooperation in politics (also called a clique) created solely for the purpose of cheating other competitors by illegitimate means? Aren't we, in the end, all selfish—even in symbioses? We will return to this topic several times in later chapters.

As just shown, if one wishes, and if it suits one's perspective, one can view nature as a system of close cooperation and collaboration. This worldview would also be very simplistic and wrong; but it is almost as widespread as the view that nature is merciless and brutal.

We can project a lot of ourselves into nature. Because we are the way we are, nature is that way too, and so it should be. We like to see in nature what we want to see. I would go so far as to say that scientists who work with symbioses have a different character than scientists who work with predator-prey systems, parasitoids, or competition. But I admit that this is hard to verify.

All of this is our projection. We see ourselves in nature; and we are all very, very different. Do some people see the brutality in nature, while others see cooperation? Does not everyone see something different in "nature"? Everyone is touched by things and events that affect them personally or remind them of something. Everyone draws different conclusions from this.

The same that applies to nature also applies to our environment in general: if we see our fellow human beings, the environment, our surroundings, etc. as evil, annoying, and stupid, this reveals much more about ourselves than we might like to admit. If we see the environment or our fellow humans as benevolent, positive, and trustworthy, we also reveal much about ourselves (which is why most voters do not like loud, angry politicians who blame external factors, because many people (unconsciously) know that they are seeing the politician's inner self).

What follows from all these processes in nature? Is nature brutal, evil, and devious (plants are not really meant here, except for the carnivorous ones…)? Is nature harmonious and characterized by cooperation for mutual benefit? Is nature diverse and do we merely see ourselves reflected in it? Is it not ultimately our own character that we see in nature? Someone who is fascinated by dinosaurs, spiders, and other predators may have a different character than someone who is enthusiastic about symbioses, cooperation, and Gaia in the realm of organisms. Do zoologists and botanists have different characters (hard to verify, but not impossible)? Is any particular view of nature better or worse?

When any conclusions are drawn from all these considerations about nature, this is called biologism. This means that we derive a normative state (an OUGHT) from what we find in nature as a factual state (an IS)—for example: nature is brutal and evil, therefore we can, may, and should also be that way (survival of the worst beast; social Darwinism), or: nature is symbiotic, harmonious, good, and cooperative, and therefore we can, may, and should also be that way.

Ultimately, however, there is absolutely no reason or justification to infer any normative state from the factual state of nature. Regardless of the fact that there is no uniform factual state anyway (nature good or bad?), as humans we are free to behave as we wish. We have a reasonably free will and can determine our actions ourselves. The accusation of biologism is therefore only justified if we try to derive something from nature.

So, in nature there is predation, parasitism, and cooperation, and we can project anything we want into nature. However, all these projections are nonsense, and ultimately reveal more about ourselves than about nature. If you look online under "what can we learn from nature," you will find many bizarre ideas and suggestions, but they all lead nowhere, because any derivation and justification of human behavior based on processes in nature is pointless and prevents any further development.

If the factual state had always been the normative state for the future, there would never have been any progress—neither in evolution nor in human history. If everything had always remained as it was, we would still have slavery, no women's rights, and there would be child labor, etc. Therefore, the only sensible conclusion, if we want to read anything from nature—or project anything into it—is that it is neither possible nor meaningful to do so. From the factual state of nature, the following applies to us humans: NOTHING.

At this point, this chapter could end; most people would probably agree with me and the topic would be settled. But a bad feeling remains. Does this also mean that the "nature" of humans is completely irrelevant for "morality"? Are we detached from biology? Do we float above it? Can we act completely freely? Here, too, not everyone would necessarily agree, and would thus judge in contradiction to their agreement above… either something follows from the factual state (biologism) or it does not. How do we get out of this dilemma?

Presumably, we need to soften our position somewhat (i.e., biologism is evil)…. To get started, I'll borrow one of the most famous quotes from Bert Brecht (a writer and theater director above suspicion of biologism): "First comes food, then comes morality." There is hardly a thought more biological than this statement. First comes the provision of resources, then come culture, literature, the arts, and morality. Thus, morality comes after resource provision (even during the Corona crisis of 2019/2021, supermarkets were not closed; but cultural institutions were). With this, Brecht—somewhat resignedly—wants to tell us that, of course, basic needs must be satisfied before we can think about ethics and the like. His choice of words is also somewhat polemical. He uses the word "fressen" (to feed, as animals do) and places us in the animal kingdom, because there, creatures "fressen." We humans usually try to distinguish ourselves with the word "essen" (to eat). Brecht makes it clear once again that it is an illusion to believe that we humans fundamentally set ourselves apart from animals. It becomes clear, for example, that basic human needs must be met before other issues can be addressed. And, of course, we need food today, or at the latest in the next few days, otherwise the most important thing we all have—our lives—comes to an end.

This simple conclusion has major implications for the crises currently facing humanity. From the spread of viruses due to high population densities to the protection of nature (in our own interest and that of our (potential) children; not even for nature as such). In all these crises, it is often overlooked that, while it makes sense to restrict contact (to contain Corona), we also know that supermarkets must remain open, otherwise there would be riots. The same applies to the issue of nature conservation. Of course, we all basically agree that it makes sense to stop using fossil fuels (since they are a limited resource and release CO_2, including the greenhouse effect), but we must never forget that this noble goal of sustainability sometimes comes into conflict with the concrete goal of feeding my family and myself. Of course, it is easy for a member of parliament, a civil servant, or an employee to demand all these things, but if I work in the coal or oil industry, I may be a bit more hesitant when it comes to implementing environmental regulations.

This issue even reveals a particular problem of democracies. While dictatorships or autocracies like China or Russia are able to implement any idea from the top down immediately and without opposition, democrats are always concerned that, although they may be doing the right thing in the long term, they might unfortunately be voted out in the short term because their strategy was too quick for people in the here and now in terms of resource acquisition. This problem cannot and must not be ignored by nature conservation. In democracies, the population must, to a large extent, at least agree with the ideas of those in power (and vice versa). This issue will become apparent again when the Corona topic fades into the background and global and local nature and species conservation return to the spotlight.

The resource situation of people also plays a decisive role in world politics. We (Europeans), in contrast to dictatorships and autocracies, define ourselves largely by human rights, ethics, and similar principles—and rightly so! It is our only unique selling point and of crucial importance for group conflicts in the future, as the significance of religions continues to decline; but what happens if, in our focus on human rights, ethics, and nature conservation, we neglect our resource (the economy)? At the moment, major migration flows, driven by a lack of resources, are heading to the USA and Europe, because there are resources there AND human rights and nature conservation efforts; but what happens if Europe and the USA slowly lose their economic dominance, especially to, e.g., China? The USA is already financially dependent on China (e.g., due to its high mountain of debt); Europe also gets most of its goods from China. China is the largest trading power; China influences the entire world, directs the flow of goods in Africa, and in Asia via the new Silk Road. Anyone who objects is disciplined. If this dominance continues to grow and jobs in Europe and the USA decrease (would decrease), there will be new migration movements, e.g., to China or other growth markets. Where I can feed myself and my family, I will secure my work and home (maybe not in the next generation, but in two to three generations). The human rights situation and the surveillance state then come second on the list of priorities. So if the West wants to compete with the new superpower China, it must pay attention to everything: the economy, prosperity, human rights, and nature conservation. Only this combination will make it possible to create a secure and free future in prosperity. Otherwise, China's prosperity, together with its autocratic ideas, will spread worldwide and displace our liberal democracies.

If we return to the topic of ethics and the question of what nature has to do with ethical issues, one thing becomes clear from all these considerations. We should never infer any kind of OUGHT for humanity from the current state of nature; but we must also never ignore human nature (as a biological being) if we want to engage with morality and ethics! There is no human ethics that exists independently of our nature as biological beings. We are part of nature and in no way separate from it. We are experiencing this right now in crises such as Corona or the worldwide decline in species numbers and global temperature changes. So what can we do? Are biologists allowed to contribute to this topic?

Let us take a closer look at one of these problems: the decline in species is a worldwide problem that we are currently facing. The numbers are usually sobering. According to the "Living Planet" Index, vertebrate populations declined by 68% between 1970 and 2017, and this trend does not necessarily stop with invertebrates. Surely everyone has heard of "save the bees" and the associated campaigns against insect decline. The causes of this decline are advancing habitat destruction—such as the clearing of rainforest areas, environmental pollution, climate change, or the excessive exploitation of animal populations—such as overfishing. The numbers are shocking, but what are the consequences? Why are we as humans affected when insects go extinct?

This question can be plausibly answered with ecology: if we continue to lose species or if population densities of species decline, natural systems become increasingly unstable. Relationships between populations inhabiting the same area can be disrupted; for example, fewer plants are pollinated if the density of pollinating insects decreases. Other types of relationships, such as predator-prey relationships, can also be disrupted by the decline of one of the involved populations. If populations or even species go extinct, the network-like structure of interactions between the species in a system becomes increasingly sparse. The system as a whole becomes more unstable in the face of disturbances. If too many species, or species with important ecosystem functions (so-called keystone species), are removed from the system, in the worst case, the entire ecosystem can collapse. An example of this would be the prairies of the North American continent. With the extinction of the North American bison, the prairies lost their most important ecosystem engineer, after which large parts of this system could no longer sustain themselves. Today, there are virtually no large-scale prairies left. This decline in species also has consequences for humans.

The facts mentioned above can primarily be provided by biologists, who analyze the current state and develop models to estimate how observed trends might continue in the future. But if one is looking for a basis for moral argumentation—for example, on the question of whether it is morally right to engage in a certain form of nature conservation to curb the problem of species extinction—the consequences and the numbers are only one component. Of course, these facts are helpful, but any argument of this kind also needs a moral foundation, and this does not come from biology, but from ethics, which, in contrast to biology, is a normative science. So, to answer the question of how a particular situation should be (e.g., should we engage in interventionist nature conservation to maximize diversity?), the facts alone are not enough; they say nothing about how a situation should be. To assess the question of the ought, the aforementioned ethical foundation is needed. We can therefore use the facts to make certain decisions on an ethical basis and with the inclusion of moral knowledge, in which biological facts may be relevant; but which facts are relevant depends on the moral standpoint, and the facts alone are never sufficient to weigh a question of morality.

If one answers moral questions about nature in this way, the question of universality arises. If a committee discussing nature conservation or other ethical issues—whose relevant facts can be found in biology—reaches a conclusion, it may still be the case that not everyone agrees with this outcome. A person who struggles to provide for their family or does not feel secure in their situation will certainly not be as enthusiastic about the idea of a new environmental campaign as someone whose resources are secure. Like animals, we have basic needs for resources, security, and reproduction, and we must not forget this. To reverse and loosely paraphrase Brecht: those who are well-off can afford to talk, can afford morality, and can develop ethics and nature conservation. Those who have nothing will not care much about all this morality and nature conservation; for they have other problems—that is, for most people, resources come first, then ethics and nature conservation.

Global Change and Politics

The largest and most significant topic in ecology at the moment is "Global Change" and all related phenomena such as global warming, rising sea levels, and the resulting altered living conditions for plants and animals. Here it becomes clear that ecology, its application, nature conservation, and politics are all closely interconnected. So what is this all about? The most important questions are: Is there a global climate change, is it caused by humans, how long have we been changing the climate, and what do we want and are able to do about it? And what are the implications for global politics? Is there anything at all that has not yet been said on this topic? Are there any surprises? Anything that hasn't already been covered in thousands of newspaper articles and publications?

In this chapter, there is not much excitement at first (which is always a good start for a text ☺), since I am not a climate change denier, but largely agree with the scientific mainstream. Climate change exists, and it is largely man-made. The following will therefore address everything related to climate change: the burning of oil, gas, and coal, the associated release of CO_2, and—currently somewhat less prominent in media coverage—methane. Furthermore, it will address the political consequences of climate change and the associated structural transformation of the energy sector.But before we turn to the topic of global change, we should remind ourselves of the principle of science. Ideally, scientists have a research question or hypothesis, for example: how does increasing nitrogen (=N) input (e.g., from agriculture; currently about 40 kg N/y/ha) affect the soil fauna of beech forests? That would be a typical question from my field of soil animal ecology. To address this question, I would conduct an experiment in which I fertilize various beech forests with nitrogen (in different forms and possibly at different dosages) and measure what happens to the density and community structure of the soil fauna over the next few years. Of course, I always have a control that I do

M. Maraun, *Everything is Ecology*, https://doi.org/10.1007/978-3-658-50207-2_4

not fertilize. The crucial point is that I have repetitions, i.e., replicates. Only with replicates can I ensure that my result is generally valid.

The number of replicates is debatable: two, three, five, or more? There should be at least three or four, so that outliers (unusual measurement results) can be easily identified. With only two replicates, it is impossible to determine which is the outlier if an unusual result occurs. With repeated measurements, more really does help. However, I should not take all my replicates from a single forest, because I might have randomly chosen an untypical beech forest (which might, for example, be growing on a rare gold vein or on a former landfill); and then I cannot say anything about the effect of N addition on beech forests in Germany or Europe. So I need replicates that are far apart and thus independent of each other. Only if my samples are truly independent—which I usually achieve by having a large distance between study sites—can I make generally valid statements. When scientific hypotheses are investigated, the replicate is the most important point. If I accuse someone of being unscientific, the best way to do so is to point out that they have not investigated real (i.e., independent) replicates. The absence of replicates is a cardinal sin in science.

However, there are many sciences that have difficulties with replicates. In evolutionary biology, I cannot run evolution multiple times (except perhaps with bacteria in a test tube); in astronomy, I cannot replay the time since the Big Bang in replicates, because there is only one Earth and, moreover, only one universe (so the universe itself is without replicates; theories about multiple parallel universes we will leave aside for now); history also has this problem (there are no repetitions; "how would history have unfolded if one of the assassination attempts on Hitler had succeeded" is just a thought experiment); in economics and law, experiments with replicates can also only be conducted to a very limited extent. "How does prosperity develop in comparable countries where only the economic system differs?" "How much of a deterrent is the death penalty for criminals?" Such questions are only accessible to the scientific method with replicates to a limited extent.

And even ecology sometimes has this problem. Because there is only one Earth—we have no replicates. No one knows what would have happened if humans had not evolved, if we did not release CO_2, if we all cooperated better. We also cannot repeat the history of humans and their impact on the climate of our Earth with replicates.

Sciences in which no real independent replicates can be investigated rely on extrapolations and projections from model calculations and simulations; they try to extract trends from the available data—in short, they try to look into the future. And you don't have to be Woody Allen to know that all forecasts should be treated with suspicion, especially when they concern the future. How difficult it is to make accurate predictions can be seen right now in the third wave of the Corona pandemic in Germany (which will hopefully be over by the time this book is published); different experts make different predictions; and even if these experts all make the same predictions, they do not necessarily come true. Predictions that have not come true include, for example, the collapse of the world due to overpopulation (Hoimar von Ditfurth; Club of Rome), the extinction of Germans due to low sperm quality (which was once reported in the news), the disappearance of the German forest already mentioned in the previous chapter (Spiegel cover story 1981 "The Forest is Dying"), the end of history (Francis Fukuyama), the end of the world due to nuclear war (whoever predicted that), and so on. Natural scientists should therefore always be very careful and cautious with predictions. If they do not come true, they appear untrustworthy; and who will believe them next time? Quite apart from that, different scientists in almost all disciplines nowadays make different predictions, which initially attract a lot of attention, but are then very rarely checked in retrospect (even though this would usually be possible). If predictions do not come true, it is simply said that the assumptions of the model were not correct or were too simplistic. I think that natural scientists should above all be humble about the limits of their field and the limits of the predictability of models, since these models may or may not be correct. Unexpected, unforeseeable events occur, such as wars, natural disasters, inventions, improvements in quality of life. Complex systems are generally difficult to predict: the behavior of living beings, ecosystems, the stock market, human diseases, economies. There are general limits to what can be predicted (see Prologue).

Global change research is just such a field, dealing with a complex system (the entire Earth) without replicates. "Global Change" is therefore a phenomenon that is difficult to measure using the classical methods of the natural sciences. The research thus relies on trajectories, modeling, and measurements without real repetitions. All replicates we investigate on the topic of global change are so-called pseudoreplicates, i.e., they are not independent of each other, since they are taken from the same Earth. This sometimes makes it easy for deniers of global change to claim that the evidence for global change is not robust enough. Are there nevertheless findings and data that are so reliable that we can consider them trustworthy? Yes, there are many relatively robust findings that we can all trust. Many trends are so clear that, even when viewed with great caution, they can hardly be regarded as random or uncertain events. In the following, I will present some of the most important phenomena that can be regarded as relatively certain consequences of global change and global warming.

One example is the global rise in sea level (by about 20 cm in the last 150 years). While one can argue about the details (exactly how many centimeters?), there is little doubt about the phenomenon itself. The cause is certainly a global warming of the Earth due to an increase in CO_2 in the atmosphere and the associated greenhouse effect. As the Earth warms, glaciers and ice sheets melt, and water expands at higher temperatures, so that we are currently seeing a rise in global sea levels.

Is this caused by humans? There is certainly little doubt that the rise in sea level is associated with human activity and has also been causally brought about by it. Is this really proven? No; nothing is ever proven in the natural sciences (proofs can only be established in mathematics); but there is very strong evidence for something when the data is clear. And in the case of human-caused sea level rise, the evidence for its influence is overwhelmingly strong. Furthermore, the warming of the oceans by about one degree is also among the relatively reliable data sets. This increase in temperature often leads to the well-known phenomenon of coral bleaching, i.e., the corals expel their symbiotic algae.

If we look at greenhouse gases, in addition to CO_2, methane is also a greenhouse gas that is (indirectly) released into the atmosphere by humans. Methane has almost disappeared from the current discussion; but that could quickly change again (that would be a prediction of mine). Methane is mainly released by methanogenic archaea bacteria, which are active, for example, in the stomachs of cattle: each cow (which we raise for our hamburger production) emits about 200 liters of methane per day. In addition to cattle, wet rice cultivation is also a major source of methane.

Other phenomena associated with the warming of the Earth (and thus with CO_2) include the migration of animals to cooler regions (where possible), the movement of animals and plants to higher elevations in the mountains, the general warming of the Arctic, the melting of ice there, and the resulting development of new trade routes such as the Northwest Passage north of Canada and the Northeast Passage north of Russia. Ships to and from Asia could travel these routes as an alternative to the Suez Canal. This is also linked to the construction of new ports, which of course has great political significance.

Furthermore, the Antarctic ice shelves are melting; glaciers are retreating worldwide, the Earth's permafrost soils are thawing, which can lead to damage to buildings. In addition, there may be further release of CO_2 from these soils, accompanied by a positive (for us humans, negative) feedback loop of CO_2 release. Especially from the cold biomes of this Earth, a lot of CO_2 can be released, which is due to the fact that a great deal of dead organic material accumulates there. Because photosynthesis is a biochemical process, it is not as strongly influenced by temperature, so it continues even at cold temperatures—albeit more slowly—while microbial decomposition processes (mineralization) are more strongly affected by temperature. Therefore, in cold biomes, dead organic material accumulates, which is not broken down by microbial decomposition. With global warming, the CO_2 that has accumulated in the form of dead organic material is massively released. Even in cold tropical biomes such as montane rainforests and in systems where decomposition is slowed by lack of oxygen (e.g., in floodplains), global warming can lead to increased CO_2 release. Further consequences of global warming include: plants advancing into more northerly regions (trees growing in what was formerly tundra); vegetation boundaries shifting. Similar things are happening with animals. We have more invasive species worldwide, mosquitoes can transmit diseases in northern latitudes that were previously unknown there (malaria, leishmaniasis), and Greenland (actually "Green Land," because it was greener during the Medieval Warm Period around the year 1000 when the Vikings settled there) is becoming habitable again (at least more so than during the cold periods after 1350, when even the Vikings abandoned the settlement of Greenland).

That more CO_2 is also associated with higher temperatures is not doubted by any reasonable scientist. CO_2 concentrations and Earth's temperatures have always been closely linked over the past 500 million years. The reason is simple and is called the greenhouse effect: when sunlight hits the Earth, the Earth emits infrared radiation into space. Some of this heat radiation is trapped by CO_2 and other greenhouse gases and is then sent back to the Earth's surface in the "greenhouse Earth." The CO_2 content of the Earth has fluctuated greatly over millions of years of evolution, even without humans, due to various causes, some of which are not yet fully understood.

So far, the Earth's air temperature (compared to the long-term average from 1901 to 2000) has increased by about one degree. The political goal is to limit warming to two degrees (recently, 1.5 degrees has also been mentioned as a target). To get a sense of the significance of temperature fluctuations, consider the Little Ice Age in Europe, which lasted from about 1650 to 1850 and was associated with a temperature decrease of about one degree, leading to significant deterioration in people's living conditions. In minor warm periods such as during Roman times and the Middle Ages (around 1000), temperatures were about one degree higher than today, which also had a major impact on people's lives. According to IPCC (Intergovernmental Panel on Climate Change) calculations, however, future warming of up to five degrees is expected on this Earth. This would usher in a warm period in Europe and the world that has not occurred since the Eemian (about 120,000 years ago). During the Eemian, it was between two and five degrees warmer in Europe than today. The Eemian was the last warm period before the present one and was characterized by drought, sandstorms, and fire. Sea levels were significantly higher than today, and many areas that are now habitable were then flooded. Forests gave way to steppe vegetation, and forest elephants, forest rhinoceroses, cave lions, and Neanderthals inhabited Europe. Hippopotamuses lived in the rivers. It is almost unimaginable what would happen if such warming were to affect Europe and the Earth. Agriculture would no longer be possible in large areas. Irrigating such vast dry steppe areas is almost inconceivable. This would certainly be accompanied by migration movements of people and massive political changes.

A seemingly simple question in between: Since when have we been changing the climate? Since industrialization 150 years ago? Since the beginning of agriculture 10,000 years ago? Since the evolution of modern *Homo sapiens* about 200,000 years ago?

Let us begin with our influence on the climate. There is little doubt that we have been influencing the climate since industrialization 150 years ago. I already learned in school about the linear increase in CO_2 levels (measured on the Hawaiian island of Molokai), which is caused by the release of CO_2 by humans. This curve is a linearly rising zigzag curve. The zigzag pattern arises because in spring more CO_2 is fixed by plants, while in autumn and winter more CO_2 is released (caused by microbial respiration of soils). The CO_2 measuring station was deliberately set up far from human influences on the remote island of Molokai to exclude local fluctuations in CO_2 emissions caused by humans.

But haven't we been influencing the climate for a long time already? For 10,000 years, since the end of the last ice age, humans have been massively altering the vegetation of Europe and other regions of the world. With the advent of agriculture, there was widespread deforestation and the cultivation of crops worldwide; especially wheat in Europe and rice in Asia. This was certainly associated with a massive release of CO_2 and production of methane. In his book "Plows, Plagues, and Petroleum," William Ruddiman shows that the increase in atmospheric CO_2 and methane concentrations has been detectable for about 8,000 and 5,000 years, respectively (not as strongly as today, but it is already visible in the data). This means that humans have been influencing the climate for thousands of years, albeit not as strongly as in the last 150 years. Ruddiman even goes a step further in a thought experiment. He notes that we are living in an interglacial period, and that the next ice age should actually be imminent. However, it is possible that humans, through the release of CO_2 and methane, have counteracted the natural cooling of the Earth just before the next ice age. We may therefore have saved ourselves from the next ice age; and a new ice age would most likely have been far worse for humanity than a warm period. In warm periods, humanity has usually benefited from larger areas available for crop cultivation; and to a certain extent, such (positive) changes will also occur with slight warming of the climate in certain regions. Areas that were previously uncultivable due to short growing seasons can now be made usable. Of course, this again leads to further CO2 release and further warming. The catastrophic consequences of even a slight cooling of the Earth can be read about in Philipp Blohm's book "The World Turned Upside Down." Cooling leads to shorter growing seasons, crop failures, rivers freezing over for long periods, and ultimately to migration flows and political upheavals.

But we do not have to deal with the problems of global cooling; our current problem is the opposite. Now, global change is one of the biggest topics on the political agenda worldwide, and the question is: what do all these considerations mean for our actions? In my opinion, there is only one logical conclusion to be drawn from all considerations regarding the release of CO_2 from fossil fuels: It is irrelevant whether we have prevented the next ice age or not, and it is also irrelevant whether there has been a temperature increase of 1.1, 1.6, or 1.9 degrees so far. Just as the goal of limiting global warming to two degrees is a completely arbitrary political boundary. The 2-degree target was officially recognized at the UN Climate Conference in Cancún in December 2010. Whether a tipping point is reached at a global temperature increase of more than 2 degrees, at which the Earth becomes uninhabitable, is scientifically disputed and generally unlikely. When the 2-degree limit is exceeded in a few years, politics, with its characteristic cynicism, will probably set the new "target" at 2.5 degrees (even if the Earth simply refuses to comply…).

Ultimately, it is also irrelevant how precisely we scientists can demonstrate global change, and it is also irrelevant whether the vegetation of this Earth will ever be able to absorb the excess CO_2 (probably not). The only important thing is that fossil fuels (oil, coal, natural gas, etc.) are not infinitely available and their use must therefore be ended as quickly as possible. Humanity must understand that it is depriving its children of the foundations for life on this planet if it continues to use fossil fuels, because, firstly, they are finite, and secondly, they are heating up our planet in unpredictable ways.

I am not a pessimist, not a cynic, not an apologist for catastrophe, etc.: we can achieve a lot! We have cleaned up rivers (e.g., the Rhine), we have defeated diseases (smallpox, even corona by the time this book is published), we have expanded agriculture, can feed billions of people, we have eliminated acid rain, we no longer hear about the ozone hole (CFCs, where are they?), and in Europe we first created the species diversity that we now want to protect here. Initially, agriculture opened up our relatively monotonous beech monoculture in Europe and led to the immigration of hundreds of species from open habitats or riverbanks. We can also handle the CO_2 problem, but we have to recognize the problem and we have to truly want the solution. This problem is different from river pollution and other regional problems like acid rain, and after the ozone hole, it is the first global problem humanity has had to deal with. It does not matter who releases CO_2 where on this Earth; in the end, it all distributes evenly in the atmosphere.

One of the arguments of the so-called climate skeptics is still: there have always been natural fluctuations in the Earth's climate. And this argument is, of course, correct. There have always been massive temperature fluctuations on this Earth. Complex life on this planet began about 550 million years ago. Fossils from the Burgess Shale (Canada) and from Chengjiang (China) are known from this period. Before that, there were certainly bacteria, fungi, and algae, but hardly any complex multicellular life forms. Previously (in the Marinoan and Varangian; that is, at the glacial maximum about 560 to 650 million years ago), the Earth was probably almost completely covered with ice (Snowball Earth hypothesis; Kirschvink 1992); with sea ice reaching as far as the equator. After the end of Snowball Earth, during the radiation of the metazoans about 500 to 100 million years ago, there was a rise in the global climate, with temperatures up to 10 degrees (!) higher than today. Later, during the glaciation in the Permian (about 270/280 million years ago), there was an ice age with temperatures about four degrees below those of today. After the Cretaceous-Paleogene transition (extinction of the dinosaurs, etc.) about 66 million years ago, the temperature was about 10 degrees higher than today. From that point on, the climate gradually cooled over the last several million years, ending in the alternation of warm and cold periods, which we now see in the records of global temperature development.

All these fluctuations are natural; we humans had nothing to do with the emergence of the aforementioned cold and warm periods, since we have only existed as *Homo sapiens* for about 200,000 years and, moreover, did not have population densities at that time that would have been relevant for any significant impact. The causes of these fluctuations are debated: variations in the Earth's orbital parameters, such as changes in the rotation (precession) of the Earth's axis, the tilt of the axis, etc., which are described by the Milankovitch cycles, influence CO_2 release and thus the emergence and disappearance of ice ages. These cycles have intervals of about 25,000, 100,000, and 400,000 years, and are associated with the occurrence of warm and cold periods. Other reasons for natural fluctuations include, for example, solar activity or the Earth's plate tectonics (for global precipitation, it makes a big difference whether we have one large continent on

Earth or several small ones. Large continents tend to have little precipitation in their centers because water sources are far away, while small continents or islands are more likely to receive precipitation). Volcanic activity is also related to plate tectonics, which increases especially during ice ages, as land masses are pressed down by thick ice sheets resting on the continents, leading to increased volcanic activity.

Despite these many potential reasons, not all fluctuations in the Earth's climate are causally understood in their origins, which is why, for example, we cannot fully explain the Roman and Medieval Warm Periods or the Late Medieval Little Ice Age.

In summary, one can say: one can endlessly ponder the origins of natural fluctuations, and that is intellectually interesting; but it is completely irrelevant to our situation. Because the Earth is warming, and even if this were caused solely by natural climate fluctuations, we would still have to act. The fact is that we are consuming fossil fuels, which are finite, and that we must provide alternatives as quickly as possible. So we have to do something and should not just research and discuss. There is therefore hardly any reasonable doubt that there is global climate change, that humans have a significant share in it, and that we should do something about it out of self-interest. Otherwise, there will be crop failures and famines, because many areas of this Earth will no longer be suitable for agriculture.

But what if Trump and the other deniers of global change were right? Then what we measure and observe would be the result of natural fluctuations, and humans would have nothing to do with it. No one can rule this out, since the natural sciences cannot prove anything, but only provide evidence for something. However, in this case, nothing different would follow for our behavior. Our behavior should be exactly the same as in the case of human-caused global change. We should also plan for the long term, promote renewable energies, and reduce our dependence on oil, gas, and coal. Because even if climate changes were natural fluctuations and we could hardly influence them, we should still not make ourselves dependent and vulnerable to blackmail by oil- and gas-exporting countries like Saudi Arabia (and other OPEC countries) or Russia (and other gas exporters). Conservatives in particular should understand this argument.

And yet another thought experiment: What if CO_2 were not a greenhouse gas and fossil fuels would last for centuries to come? Even then, politics should act just as it is now slowly beginning to do, because humanity must move toward a sustainable use of its resources. The sooner, the better.

And one last thought experiment: what will actually happen in 2050 (the year the EU aims to be climate neutral)? What will happen to Venezuela, Saudi Arabia,

Russia, etc.? We should start thinking about this. Will they need aid programs? Will these countries be desperate, act out of desperation? Those who rely solely on oil, coal, and gas exports for their economies may soon have no income at all. How will these slower-moving countries react; will the West need to help? Once the West has switched to 100% renewable energy, the world will be a different place. Let us remember that the attacks in the USA were carried out by terrorists who were indirectly supported by wealthy countries like Saudi Arabia; such countries would then no longer have the money for such things. Let us remember that Russian expansionist efforts (Crimea, Donbass) were financed with exactly the money that the West and other countries pay for Russian gas deliveries. All of this would change. The world might become more peaceful (if all goes well). But the world could also become more difficult for countries that are too slow to focus on renewable energy, or that depend on the export of fossil fuels (Venezuela, Russia). The fact is that the entire world must end its consumption of fossil fuels; China and Russia should also switch to renewables in their own countries and refrain from building coal-fired power plants in Africa or elsewhere. 'By the way,' we also need to replace nuclear energy, since no one can take responsibility for storing radioactive waste on our planet for millions of years. So we need a rapid and efficient transition to renewable energy (in Germany, especially wind; since hydro and photovoltaics do not contribute much; in other countries this may be different), and as humanity we can achieve this, just as we have overcome many crises on this earth, but it will require a gigantic effort, and this effort must be global.

That would be a nice closing statement, and most people would agree, but it does not seem that our actions go beyond lip service. Why is that? Why is environmental protection actually so difficult? Greta Thunberg is very clear when she says: Here is the problem, this many gigatons of CO_2 are in the atmosphere; it can be solved in this and that way; no more lignite, no more hard coal, no oil, no gas; yet it does not seem to be that simple. Apparently, we do not have a problem of knowledge, but a problem of implementation. And this problem is the tragedy of the commons. A resource that is available to all (the environment) is also used by all. No one wants to pay for it, and it is always someone else's fault when this resource is depleted. The blame of others and the tragedy of the commons are what we will examine in the next chapter.

'Tragedy of the Commons' and Altruism

At heart, I am a liberal. For me, this means that the individual should have priority over higher levels such as family, state, and society, and that the individual should be protected from encroachments by superior authorities. I have always had difficulties with higher authorities and hierarchies. As a child, I never understood why one should do something without reason, just because "one" does it or doesn't do it. I questioned all rules, and I only wanted to accept what I considered reasonable (e.g., brushing teeth was OK; dressing up nicely because we were visiting grandma seemed unnecessary; religions always appeared completely pointless to me, especially in the post-Enlightenment era). At school, I never liked reading books assigned by teachers (I had to read "Der Untertan" by Heinrich Mann, but would have preferred to read "The Catcher in the Rye" by J.D. Salinger or "Steppenwolf" by Hermann Hesse); I hardly participated anymore (except in art, sports, biology, and civics, because I found those exciting), and had to repeat the 12th grade. With great effort, I finally managed to get my Abitur; what motivated me most was the prospect of studying biology.

I also found this line of thought (the priority of the individual over the group) completely logical. The exaggerated importance of higher authorities has always led to the oppression of the individual, to senseless rules, to abuses by authorities, to the oppression of women, to the surveillance of citizens, to the regulation of matters that should not concern the state or families.

During a brief interlude (compulsory military service in the Bundeswehr, then still 12 months), I slowly began to realize how little power an individual has when acting alone. An individual cannot establish a state, cannot build an army, can only acquire local resources, and cannot prevail when other individuals band together into powerful groups (states, religions, armies). An individual is relatively powerless. It can rarely resist or even prevail against group interests.

M. Maraun, *Everything is Ecology*, https://doi.org/10.1007/978-3-658-50207-2_5

I gradually realized that giving priority to the individual brings many problems. Ruthlessness toward the group by individuals, the formation of monopolies, and injustice can prevail. Individuals can join with other individuals and then enforce their interests against minorities. Groups are stronger than individuals.

The problem of the great importance of individuals or individual interests is very evident in ecology and nature conservation. Nature conservation is, in principle, an interest that we should represent as a group (we cannot locally limit CO_2 emissions; CO_2 spreads worldwide). To practice nature conservation, the whole group (i.e., humanity) must agree that this is a good idea, that everyone must spend money on it, limit their use of resources, and that everyone must make sacrifices (for the good of all; "In Publica Commoda," as the slogan of the University of Göttingen, my alma mater, goes). Unfortunately, this is incredibly difficult to implement, because the individual (usually) suffers under such regulations and does not see why they should hold back. Most would say (even if not publicly), "Why should I make a sacrifice; let others do it." Or: I will use this resource as long as no one stops me; I waste raw materials or irreversibly burn fossil fuels, I produce nuclear waste and ignore the fact that the waste will remain radioactive for millions of years and that even a democratically elected government cannot reverse such a decision. But such a view is not so hard to understand! After all: Why should I preserve this planet for my grandchildren if I do not even have enough resources to feed my own children?

Ecologists and economists call this principle the tragedy of the commons. It is the principle of egoism, meaning that I benefit from the consumption of common goods that (should) belong to no one (such as air, water, soil, fossil resources), and as a consumer I profit from this, while the group suffers (and in the end, I may also suffer, but that is a long way off…). The term "tragedy of the commons" comes from the Middle Ages and meant that herders or farmers jointly used land (pastures or meadows; in Old German, the "Allmende"). Naturally, it pays off for the individual to ruthlessly exploit this resource to acquire more resources than others, which ultimately leads to no resources being available for anyone (which can then be seen as a tragedy).

To solve the tragedy of the commons in relation to nature conservation, we therefore (unfortunately!) need higher authorities that prevent precisely this egoism. Ownership and responsible individual action are fundamentally positive and sensible; but when it comes to controlling the use of common goods, we need higher authorities to defend these common goods against self-interested individuals. Otherwise, our very natural tendency toward egoism and nepotism (the preference for relatives and often also for partners) will lead to the exploitation of common resources. Advocating for such a higher authority (i.e., for a strong state) is unpopular. The Greens are always accused of being a party of prohibitions. And they are; but for the future good of all.

The group must therefore protect the individual and our future from the egoism and ruthlessness of the respective individuals, since this egoism is in our genes. Those who put their own interests and those of their children before the common good have certainly had fewer problems with resource supply than someone who acted altruistically. Those who favor their relatives and acquaintances in resource distribution practice nepotism and kin selection; a well-known and successful principle in evolutionary biology and biology. If we do not actively prevent it, there will be kings and emperors who want to pass their power on to their children and relatives, and dictators who likewise prefer their children and grandchildren as successors. That is why we need higher authorities to prevent the egoism of group members. Some individuals sometimes disparage such higher authorities as the "deep state," which clearly shows the rejection of regulation by individuals. Since higher authorities naturally have a lot of (control) power, they must not operate without oversight. Power will always be abused if it is not actively prevented.

The problem of (apparent) egoism ("America first") can currently also be seen in the USA. Many people choose the seemingly short-term improvement of their living conditions (through protectionism against Europe and Asia); in the long term, however, this could become a problem if other countries also start imposing tariffs and closing off their countries.

The same applies to Brexit. There, the population believed the demagogues who claimed that it would be worthwhile to leave the EU. So far, enthusiasm for Brexit among the population has been limited. In the long run, the UK will probably rejoin the EU.

The problem was also evident during the Corona crisis. Individuals must restrict themselves so that the group (i.e., all individuals) is protected (whereas a vaccination is not a major restriction of life; unless one believes the conspiracy theorists who claim that vaccinations also increase the incidence of autism). How difficult this is for many people was seen in the demonstrations, in the anger and despair that some of us feel when they are patronized and regulated. Here, too, there must be an authority that protects the individuals in the group from the selfish interests of a few (who do not follow any rules, but are still protected by the herd immunity of the group).

Ultimately, the tension between the (short-term) rights and interests of individuals and the (long-term) interests of the group in modern states (and in general) will never be completely resolved; there will always be dynamics at play, sometimes one side will prevail, sometimes the other. In very individualistic countries, it is certainly sometimes a great challenge to implement group interests, while in countries where the group interest is placed above the individual interest, it leads to surveillance and oppression of individual members.

Cultures in which the individual is less important than the group, such as some Asian countries, are strong groups. They do not allow individual selfishness and can (or at least: could) implement ecological and economic group goals without much opposition. The problem, of course, is that countries with powerful overarching governments tend to override individuals, spy on them, and punish them for having their own ideas and projects. The question for liberal Western democracies now is: how do we deal with such autocracies and dictatorships? Do we cooperate with them, since they are important export markets for us, do we keep a critical distance and occasionally point out individual rights, or do we become more and more like them (which would be the dream of autocrats), because then we too could make decisions for the group without having to conduct a democratic discourse on every issue. The problem with autocracies and dictatorships, however, is always that only a few individuals make important decisions, and these decisions are not necessarily better than those reached after a long and intensive democratic discourse (whether I am right about this, history will tell).

Ultimately, one must ask: does the West benefit from globalization (and many Western countries are indeed strong exporters), or does the rest of the world benefit more from globalization (due to a leveling of wages and living conditions)? If you look at the trend in Western countries, where it is becoming increasingly difficult for a skilled worker to support a family with two children, leading to tendencies toward isolationism and protectionism everywhere, one can rather conclude that the world benefits from the equalization of living conditions and that the wealthier countries will have to accept a loss of prosperity.

A few years ago, 'attac' (Association pour la Taxation des Transactions financières et pour l'Action Citoyenne) was a relatively well-known organization known for its criticism of globalization (currently, it no longer plays a major role in public perception). But attac illustrates the problematic nature of the globalization issue. On the one hand, attac was and is against (uncontrolled) globalization, as this supposedly only benefits large corporations; on the other hand, attac also recognizes that globalization actually distributes wealth more evenly, which leads to a loss of prosperity in the so-called Western countries. Thus, attac represents, on the one hand, global group interests (worldwide justice, high environmental standards) and, on the other hand, tries to represent the long-term interests of individuals. In fact, however, attac's agenda means that the standard of living in the wealthy Western democracies would have to decline (so that there is global justice and a fair distribution of resources). Not everyone is happy about that.

So if things become fairer in the long run, there will be a decline in the standard of living in the West. It will no longer be so easy to demonstrate against 'global change' in the morning and fly to Thailand in the afternoon, as was recently the case with some climate activists.

Ultimately, we see in elections in Europe and the USA that the trend toward selfishness and isolation is increasing; all of this is a consequence of globalization. We like to think about the interests of the generation after next and of "others" when we are doing well, but in times of crisis, our true preferences become apparent. In short: when we are doing well, we vote left and liberal; when we are doing badly, we vote right and reactionary.

The tragedy of the commons thus shows that we are selfish and that the tendency toward nepotism is very strong. Therefore, it will always be necessary for higher-level authorities to keep individual selfishness in check.

Worse than the tragedy of the commons is only the so-called "normative power of the factual," which states that when something is irrevocably gone or when something irreversible has happened, then it simply cannot be changed anymore. In biology, species can go extinct; in ethnology, there are peoples who have died out (e.g., Inca, Maya, Etruscans, Sumerians, Neanderthals); in politics, there are countries that no longer exist and decisions that are irreversible (e.g., the establishment of nuclear power, which leaves us with radioactive waste for millions of years), and everyone knows that there is no turning back when someone has been murdered. The partial occupation of Ukraine by Russia also belongs to the things that fall under the "normative power of the factual." Even if peace negotiations between Russia and Ukraine were successful, it seems very unlikely that Russia would ever leave the occupied territories. To mitigate the "normative power of the factual," we also need higher-level authorities such as courts and the police to prevent such irreversible decisions. However, such higher-level authorities often do not function well, as they can only sanction decisions after the fact, and they themselves are not without self-interest. Groups generally only work well together when it comes to acting against other groups. That is why the UN does not work when there are conflicts on this earth. Only if we had a common enemy on earth (a hypothetical attack by aliens from space on earth) would the UN work, because then everyone could and would have to unite against a common enemy. If we then defeated the aliens, individual interests would again dominate.

Are we generally even capable of pursuing overarching and long-term group interests, or will selfishness always prevail? Can nature conservation ever succeed if it requires sacrifices from individual people?

Does altruism even exist, that is, behavior that is to my own disadvantage but benefits others or the group? Of course, the group also includes my own relatives and descendants. Is it then really altruism, or is it also selfishness? Or is it as the English saying goes: "Nice guys finish last"? We should be clear that, of course, we want to preserve this planet for our descendants as well; and thus, nature conservation is also selfishness and involves kin selection; however, the problem of the difference between short- and long-term goals has already been addressed above. So, may we hope for altruism, or should it be made clear that nature conservation is ultimately also selfishness?

I'll keep it short. I'm sorry, but there is no altruism—even though it often appears that way. I will go into more detail on this below. So, when it comes to nature and environmental protection, we should not rely on people aligning their actions with the interests of others, but always act selfishly. This does not mean that one should be pessimistic, but only that one should be realistic.

Altruism, that is, behavior to one's own disadvantage and for the benefit of others, is often confused with many different behaviors.

Altruism can easily be confused with kin selection; however, it quickly becomes clear that helping and supporting relatives is, of course, selection at the level of genes and is therefore definitely also selfish. Even if an organism were to sacrifice itself for another, this does not have to be altruism. For example, bacteria infected by bacteriophages (i.e., viruses that can infect and kill bacteria) commit suicide to protect other bacteria from further infection. But since these other bacteria are often clonally identical to them, this suicide is also selfish. In general, with genetically identical cells, one cannot speak of sacrificing oneself for other genetically identical cells, since they possess the same genetic material. Our body cells also do not sacrifice themselves for our germline (which alone is passed on to the next generation), but support the survival of the germline. The DNA of the germline is identical to that of the body cells (at least before meiosis), so the body cells are not really sacrificing themselves. The famous "helpers at the nest" in birds are also not an example of true altruism, since they most often help closely related individuals raise offspring.

Even the renunciation of one's own reproduction, for example in bees and ants, is not true altruism. At the beginning of the establishment of this mechanism in evolution, the workers were closely related to the queen and therefore refrained from reproduction (although the queen also made sure that no workers laid eggs); later in evolution, as the colonies became larger (up to 50,000 individuals in a beehive), the queen mated several times, thereby reducing the relatedness among the workers. As a result, the workers no longer allowed each other to have offspring; "worker policing" occurred, i.e., the workers eliminated eggs laid by other workers. This allowed the queen to maintain control over the large colonies. So what today appears to us as altruism in large hymenopteran colonies (everyone seems to be working for the good of the group) is in fact a forced cooperation in which the queen is in control.

Sometimes there also seem to be cases where cooperation occurs beyond biological relatedness. Vampire bats constantly need a small amount of blood; and if they do not get it, they can starve. When necessary, even unrelated individuals help each other out. However, the individuals know exactly whom they have helped with blood, and expect a return favor when needed, usually also in the form of blood. This mutual form of assistance, which is independent of relatedness, is called "reciprocal altruism"; it has nothing to do with true altruism.

Even if I let a car merge into traffic (even though I have the right of way), that is not altruism, because I hope to be let in on another occasion as well. This is called indirect reciprocal altruism.

Cooperation that looks like altruism could also evolve via a so-called "green beard." This is any visible trait (for example, a hypothetical green beard) by which carriers of the allele/gene for this trait recognize each other and provide mutual support. This may look like altruism, but on the genetic level it is also egoism.

Even "spite" could look like altruism, for example if an individual were to punish another that had behaved negatively toward a relative of the punisher or negatively toward the group norm. Such a punishing individual takes a certain risk of being injured in the process of punishing. However, this too is not genuine altruism, since in the end close relatives would benefit.

Of course, there can also be cooperation between non-related individuals (this too is not altruism); for example, in trade between countries. In this case, both trading partners are relatively cautious at the beginning of the interaction, since the possibility of cheating is relatively high. Only after trading with each other for a longer period does trust develop and the trade volume can increase.

Some might also assume that "charity" and all comparable forms of philanthropy could be something like altruism. However, only the wealthy can afford such events; so it would be more accurately classified as showing off. Evolutionary biologists call this "costly signalling"; it means that as the giver I am indicating that I am capable of surviving even if I throw money and resources out the window. Kurt Tucholsky jokingly said about this, "Do good and talk about it."

Ultimately, cooperation is always associated with gaining advantages over other individuals or other groups. For example: If I were to engage in free trade with the whole world and only one country were not included, this would certainly have negative consequences for the prosperity of that one country.

However, this conclusion also implies, conversely, that cooperation can never work if it does not unite against an outside party. And if you look at the extremely limited effectiveness of the UN (has the UN ever ended a conflict or war?), then this conclusion seems plausible. The UN seems to be the very epitome of the proverbial toothless tiger.

It is therefore important, in the context of efforts for nature conservation or generally for protection against the tragedy of the commons, to appeal to people's egoism rather than their altruism. In the end, of course, all individuals suffer from a destroyed, exploited, and toxic environment.

So what can we do? Individuals are essentially egoistic (altruists who, out of concern for the environment, did not have children are not among our ancestors); but for the long-term survival of the group (which includes the individuals of the future), there needs to be control of this egoism and even a certain degree of cooperation, and thus also a certain degree of self-restraint among the members of the groups. Squaring the circle seems easy by comparison.

We need a sense of community, but it must not (too much) come at the expense of the individual. The individual must always be the most important unit of the state/group; it must always be protected from the overwhelming interests of the general public. But sometimes the interests of the group are identical to the long-term interests of the individual, e.g. defense of the country and democracy, preservation of the environment, so that out of self-interest we should/must temporarily subordinate our egoism to the interests of the group. For example, with Corona. Only because so many people got vaccinated did the whole group ultimately achieve herd immunity (from which even the egoistic and ignorant anti-vaxxers benefited). So everyone should first be able to do what he or she wants, as long as the boundaries of others are not crossed (in a sense, Kant's categorical imperative) and as long as the group does not suffer in the long term (topic: environmental protection). Unfortunately, this is not always easy, since we are certainly often the children of people who benefited from crossing boundaries.

So resistance to the implementation of climate goals and environmental protection is to be expected for reasons of egoism. This egoism is to be understood as a result of evolution, in which there is no altruism. We must therefore make it clear that we must protect and preserve our world out of egoism. People who are against nature conservation and against stopping climate change (even though they would rarely admit it) are usually afraid of the changes involved and above all of the costs. And because that is not a nice ending to a chapter, I would like to put it the other way around and positively. For nature conservation and for avoiding the tragedy of the commons, we need a lot of scientific understanding, a lot of courage, and—despite all the problems—a lot of optimism.

Populations and Demography—Pensions, Jobs, Migration, Asylum, Growth

Populations normally grow exponentially according to their growth rate (r), unless something prevents this growth. And since no population can grow exponentially forever (otherwise the Earth would quickly be overrun by this one organism, which, incidentally, would then also have no resources left), there is always 'something' that opposes any sustained population growth. This "something" can, for example, be a shortage of food resources. As a population grows, it (usually) reduces the amount of its resource (if the resource renews quickly, the population continues to grow until the resource can no longer recover, or the population is regulated by abiotic factors; e.g., frost, heat, drought). When a population is limited by its resource, this is called "bottom-up" control. This often leads to competition between individuals of a species (intraspecific competition), at least when the utilized resource becomes scarce. Alternatively, a population can also be limited by the presence of a predator. This would mean that as the density of a population increases, the impact of consumption by a predator also increases (simply because the predator has more prey available); this is called "top-down control." A resource does not necessarily have to be food; it is also possible that other factors become scarce with increased use; for example, space is a resource for sessile animals and plants.

At the beginning of the study of population growth, so-called discrete population models were used; the question was how the number of individuals at time t+1 compares to the number at time t. For this, it was necessary to know how many births and how many deaths occurred within a certain discrete time interval.

A simple example is provided by grasshoppers. In our temperate regions, a female grasshopper lays about 200 eggs before winter, from which the larvae of the next generation hatch the following year. Only a few of the eggs actually survive to become adult animals. The mother itself does not survive the winter, so the generations do not overlap. Thus, the development of the population depends

M. Maraun, *Everything is Ecology*, https://doi.org/10.1007/978-3-658-50207-2_6

on the number of eggs that develop into females, which themselves will have offspring the next year. If you systematically compile these parameters, you get a life table to predict the current population size and its possible future development. Such a life table can, however, be much more complicated than in the grasshopper example. Often, in addition to birth and death rates, changes in the number of individuals due to immigration and emigration must also be considered. It can be even more complicated with plants. Plant seeds sometimes survive for several years in the soil (they form their own seed banks) and thus survive several winters before finding a good opportunity to develop. And it becomes even more complicated when plant or animal species live for many years and their reproduction rate changes with age (as with humans). To be able to make a prediction about population development in such cases, one would actually have to examine each cohort (age group) closely to determine their respective reproduction and death rates. Prediction is further complicated by the reproductive mode of the population, since in sexual populations only half of the individuals—namely, only the females—produce offspring. The actual calculation of a population's development thus seems hardly possible, because too many factors must be considered, and because complex systems are, of course, only predictable to a limited extent anyway.

However, over the years, population biologists noticed that a reasonably reliable prediction of the development of population density can be calculated solely based on the current density of the population and its growth rate r. All other parameters mentioned above can be ignored. Of course, such a calculation is not perfect, but it has helped to better understand population growth. Such population models are called (in contrast to the discrete population models I described above) continuous population models. A population thus grows exponentially according to its growth rate r until something opposes it (the so-called environmental resistance K). Environmental resistances can limit population growth, for example, when the food resource is so depleted that not all individuals in the population can get enough food. Ultimately, environmental resistances will limit any exponential growth of a population at some point. The terms growth rate r and carrying capacity K of the environment are also used later in the r and K concept, which we will discuss in the next chapter.

Populations with relatively low density should therefore grow rapidly, while populations near K grow only slowly or hardly at all. There are some complications to this idea, which I will address again later in the book and which demonstrate the great significance of the r and K concept. For example, the density of a population can be too low to ensure its survival (e.g., in colonial nesting birds or schooling animals; this is called the Allee effect); in such cases, the population is doomed to local extinction. K does not have to be a constant number or limit, since populations may be able to shift their K (humans do this through agriculture, allowing us to achieve a higher population density than through hunting and gathering). Furthermore, the growth rate r can also be so high that the population density overshoots the possible carrying capacity, only to drop sharply again soon after.

What follows from all these considerations for us humans? First of all, we are also organisms whose growth is limited by various environmental conditions. We do not stand outside of nature, as is sometimes claimed. We have to feed ourselves, protect ourselves from diseases, and we compete with other species and among ourselves for shared resources. Top predators that consume us are hardly a threat. The death rate from sharks, lions, and crocodiles is negligibly low for humans. Our most important causes of death are rather serious diseases such as stroke or cancer; recently, however, also viruses that benefit from our high population densities.

To understand population development and its significance for people in Germany and worldwide, one must, however, take a closer look at the respective age cohorts, especially since population development in Germany (and other Western countries) is different from that of the entire world.

If you look at Germany's demographics in 2025 on the website of the Federal Statistical Office, you will notice that it is not a homogeneous pyramid, but that there are bulges and troughs. What caused these? In the age pyramid, you can just about see among those born in 1930 and earlier that there were many deaths among men in World War II. In addition, problems with resource supply during the global economic crisis of the 1930s also negatively affected birth rates. After the end of World War II, another drop in births can be seen due to the problematic economic situation in Germany. Then, in the 1950s and 1960s, the period of the economic miracle in Germany began, which led to very high birth rates (the 'baby boomers') of the 1960s. In the 1970s, birth rates slowly declined again, which is sometimes attributed to the invention of the birth control pill for women (the "pill dip"). Regardless of whether this theory is correct, it is interesting to note that (unlike other declines in birth rates) this decline was not caused by war or poor resource conditions. Possibly, women's self-determination has outweighed the importance of resource conditions. Future research may further investigate this issue. After the reunification of the two German states, the Federal Republic of Germany and the GDR, there was another decline in births, which was probably again related to economic uncertainty in Germany (especially in eastern Germany). After that, birth rates stabilized at a level of about 700,000 per year, which is about half the birth rate of the baby boomer years (about 1.3 million per year).

If you look at the demographics a little more closely, it becomes clear that they have outstanding significance for politics. The data clearly show that in the next 10–20 years, many baby boomers will retire. This means, on the one hand, that many jobs will become available, but on the other hand, that fewer working people will have to support more retirees. This problem is, of course, further exacerbated by the increasing life expectancy of Germans, which is not yet taken into account here. This is not a topic that is easily addressed in a democracy, since the focus of politicians and voters is usually rather short-term. Long-term planning is generally not desired, especially if it has negative implications for the results of the next election. It is more likely to be promised that "pensions are secure" (Norbert Blühm), since people prefer to hear that rather than that there are demographic problems that need to be solved in the long term. Dealing with issues that lie more than four years in the future is a general problem in democracies; this includes the aforementioned demographic change, the ecological transformation of industrial society into a society with sustainable resource use, as well as the problem of nuclear waste (which will remain with us for generations, even if we have now—at least in Germany—decided against nuclear power).

From this, it is clear that ecology and demography have great political significance. In order to prevent our labor market from drying up (on which our prosperity, our resource situation, and the flourishing of our democracy depend), we need skilled workers, and since our country, due to demographics (low birth rate), cannot train enough skilled workers, it makes sense to admit skilled workers from abroad. The exact number can be a matter of political debate, but that there will be an increasing shortage of skilled workers is beyond question. Even now, it is already apparent that skilled workers are lacking in many areas of the labor market.

This is linked to another problem, or rather its possible solution. The pension level (50.5% in 2017; fixed by policy at 48% until 2025) may not be sustainable in the future, simply because too few contributors have to finance too many (and ever longer-living) retirees. This is something that politicians are reluctant to admit. This problem affects the pension system just as much as nature conservation and climate policy—issues that will occupy us for decades to come. Immigration, especially of skilled workers, will partially solve the pension problem, since a certain amount of immigration can also help secure the pension fund (or the pay-as-you-go pension system; after all, the pension is not funded, even though this is repeatedly proposed). That immigration can solve this problem is something most parties prefer not to mention, even though even the CSU would accept up to 200,000 immigrants per year, while the head of the Federal Employment Agency even recommends 400,000 immigrants per year (Detlef Scheele,

2021). I therefore think it has become obvious that ecology, resource availability, and demography are political issues, and that ecology and thus ecologists absolutely must contribute to the debate.

Future migration movements (which are foreseeable and in some cases already well underway) due to poor living conditions are, of course, also linked to the issue of asylum. So far, it is mainly those who are politically persecuted who are recognized as refugees (although this has de facto been abolished in Germany, since refugees in the EU are actually required to apply for asylum in the first country of contact) or people whose lives are threatened in their home countries for various reasons. Other refugees are classified as so-called economic migrants, a second-tier category. They are granted asylum only in rare cases. From an ecological perspective, this distinction between political refugees and economic migrants is questionable. Why should poor resource conditions and lack of prospects (which can threaten my existence and my life) not be a truly valid reason to seek a new start in another country? The desire for asylum is almost always also resource-driven. Resources are a political issue; that is why almost every election poster in Germany, alongside freedom and security, always states that the respective party wants to secure prosperity. For without prosperity (=resource), even the best political system is of no use. This is also the greatest concern of the so-called Western democracies—or at least it should be. If prosperity were at risk, and people in dictatorships like China and Russia could live better, then democracies would be in danger. That is why all Western governments try to strike a balance: on the one hand, trading and communicating with such regimes (to maintain their own prosperity), and on the other, repeatedly pointing out the problem of human rights. We should not forget that the reasons why so-called economic migrants, negatively connoted by us, seek asylum are privileges we were born with (for which we did nothing ourselves, but simply had the luck of our birthplace). Nevertheless, many Germans also emigrate to countries like the USA or Australia in the hope of prosperity, and there are no politically persecuted German citizens among them.

Therefore, a new refugee and asylum policy should be appropriate and oriented toward the interests of Germany (the West). This means that we can, in fact, allow migration. This reasonable position would lie between the positions of the AfD (no one is allowed into the country anymore, no matter how unreasonable and inhumane that is) and the positions of the political left, who would like to admit almost all refugees to Germany. Out of self-interest (shortage of skilled workers, pension problem; see above), we can take in many people, but we cannot take in everyone. It is also not acceptable for someone like Andrea Nahles to be criticized for stating a political truism (when she said that we cannot take in all the people of this earth in Germany), just because it is currently not opportune. Whether the number of admitted migrants ultimately ends up at 100,000 or 200,000 must then be part of a reasonable, rational debate and must be continually readjusted.

The situation regarding world population development looks quite different (in contrast to Germany). There, we are in the phase (or at the end?) of exponential population growth. As early as 1972, the Club of Rome warned of the strong population growth on earth and the associated overuse of the planet's resources. At that time, however, the progress of modern agriculture was not yet foreseeable, and over the decades it became clear that this earth can support a higher population density than previously assumed. To this day, this trend seems almost unbroken, and famines are not a serious issue on this earth (although there are always local resource problems). Steven Pinker is even so optimistic in his book "Enlightenment Now" that he believes that a still-growing population on this earth can be fed by modern agriculture. Apparently, as humans, we are able to shift our carrying capacity K significantly upward, i.e., we can reach previously unimaginable population densities without the resource becoming scarce. As an ecologist, however, it is relatively easy to predict the end of this process. No matter how modern and efficient our agriculture becomes, we will eventually perceive the limits to growth, i.e., we will reach K. Ultimately, there will be a regulation of population density. The simplest way would be if we were to regulate our population densities ourselves by mutual agreement. However, this would be a relatively unique process in ecology/evolutionary biology. There is no organism that self-regulates its population. The history of self-regulation of populations in ecology is a history of errors, culminating in the fabricated Walt Disney story of the self-regulation of lemming populations. In the Walt Disney film "White Wilderness" from 1958, it was allegedly shown that lemmings, at high population densities, tend to reduce their population densities themselves (in the case of overpopulation) by jumping off cliffs or drowning themselves in the sea or rivers. However, this film was not a documentary, but a staged mass suicide of lemmings, which were thrown into the water for the film. The background for the (indeed occurring) migrations of the animals and the sometimes associated deaths is simply the strong fluctuations in populations, which can be caused by years with good resource availability and few predators. The temporarily high population densities then lead to a lot of offspring, who in the following year no longer find enough resources, resulting in the well-known mass migrations with a certain mortality rate. The migrations of the lemmings are thus an indication of the individuals' will to survive (i.e., individual selection) and not an indication of mass suicide or other ways to reduce the population density for the good of all (at least the survivors). The latter would be a "group selection" argument.

We can therefore conclude that there can actually be no self-regulation of a population. Such a population would be doomed to extinction in the medium and long term, as it would lose out in competition with other populations (it is no coincidence that almost all religions say, in effect, "…and multiply diligently"). However, surprisingly, there is hope. In the highly developed, wealthy industrial nations, the birth rate is falling despite great resource availability. Many families would certainly be able to support more than just the statistical average of 1.4 children (mean for Europe). The decisive factor here could be women's self-determination. Due to the existence of various contraceptive methods and the increasing equality of women, it is now primarily women who decide on the number of children. This number is certainly lower than the "desired number" of most men. Ultimately, it seems that the reproduction rate in modern societies is no longer very closely linked to resource availability (although I would not speak of a complete decoupling here). In fact, the growth of the world population is currently declining, so that it is possible (at least that is my hope) that there will not be major conflicts over the available resources. This would be the optimistic view. The pessimistic view would be that we reach the earth's carrying capacity without having significantly slowed our population growth. Then the prediction would be that there will be conflicts over the remaining resources, and as a result, increased migration movements and wars on earth. The migration flows to the USA and Europe could then be just a foretaste of future migration movements. As you can see, ecology is always also politics.

What follows from the above? Firstly, not only can we cope with a certain number of refugees, we actually need them in order to stabilize our pension system; so we can be glad when people from all over the world come to us. We can also hope that, for example, after the end of the wars in Ukraine, Syria, and Israel, not everyone will return to rebuild their countries, but that some will stay in Germany. The number of refugees we ultimately take in is then a political decision. We should put this to a vote. However, we should recognize that we are a country of immigration in the heart of Europe; and we should respond to this with a modern migration policy. We need qualified professionals, but we can also train many young people ourselves. We can be grateful that so many people seek refuge with us from resource scarcity (and of course also from political persecution), who in turn help stabilize our pension system. So, in a way, it is quite selfish of us to help.

r- and K-strategists—Everything is Lobbying—Even the Desire for Peace

Of course, there is no objective morality. At least not one you can pick up somewhere, like Moses did with the Ten Commandments. Neither nature nor ecology helps us much further, either. Yet the longing for the security of some external aid in dealing with small and large moral issues is very strong. Sometimes our longing even seems almost infinitely great. We would like to have a moral fixed point. After all, there is no doubt that we are social beings, and as such, we seek rules for living together. We seek meaning, a sense of community, norms, orientation, rituals. Ideally, even universal values. Here, I would like to show that we can derive such rules from nature only to a very limited extent, that we can nevertheless learn from nature, and how cautious we should be with general conclusions in the field of ethics.

Apparently, it is extremely difficult to find universally valid rules for human coexistence that are accepted by everyone. Throughout human history, many different sets of rules have been established, but most of them lasted for varying lengths of time. An important cornerstone of morality (at least in the last thousands of years of human history) has been religion. Religions have provided and still provide us with apparent instructions for proper behavior. Unfortunately, different religions give different directives for action, so it does not seem as if we will ever find a universally valid, universal morality. That is, of course, frustrating. Religions are also increasingly losing support among the population, as can be seen in Germany by the declining number of members in the major churches. In Germany, about 40% of all people have no religious affiliation, and how many of the remaining 60% are truly religious is another question. Religions, therefore, will not be able to provide us with rules for living together in a permanent and long-term way. Of course, one could accuse the non-denominational and atheists of having no morality or ethics at all; but I am not aware that the proportion of

M. Maraun, *Everything is Ecology*, https://doi.org/10.1007/978-3-658-50207-2_7

non-denominational people in prison is higher than that of those with a denomination. So it seems that an ethics beyond religions is also possible.

In the end, can the natural sciences possibly help after all? Can there really be a morality beyond religion? About 10 billion people live on this earth, many of whom are organized in religions. And the others sometimes belong to sects (the difference between a religion and a sect is really just the number of members) or believe in various other things. In addition to the large, monotheistic religions (Christianity, Islam, Judaism), there are many others (Moon, Scientology, Buddhism, Hinduism, Shintoism, Jainism, Sikhism, Confucianism, Taoism, Zen Buddhism, Rastafari…) as well as various mythologies (the difference between religions and mythologies is also just a semantic one). In the past, there were also several polytheistic religions (in ancient Rome, ancient Greece, ancient Egypt). In addition, today there are many people who are not (or no longer) organized in religions and/or are atheists, or agnostics who are not quite sure about the existence of God or gods. So can any religion, or even an atheist or an ecologist, help us in the search for a universally valid ethics? For this, we must briefly return to biology and ecology.

In ecology, there is the concept of r and K strategists. It is easy to understand. For this, we need to recall the growth of populations (see Chap. "Abiotik, Biotik und Ressource"). All that is important is to remember that r is the growth rate of the population (the larger r is, the faster the population grows), and that K is the maximum capacity of the environment that a population can sustain. You do not necessarily have to imagine this K limit as a concrete ceiling that a population can never break through with its density. By manipulating the environment, for example by providing more food through agriculture, a population can significantly increase its K. We humans are the prime example of how K can be shifted far upwards. In parallel to us (but for much longer), some species of ants also practice agriculture by cultivating fungi. K, the capacity of the environment, can also be temporarily exceeded. Then the population consumes more resources than can regrow. This leads to growth beyond K. In the next generation, the population will then fall far below K, because there are hardly any resources left. This results in regular or irregular population fluctuations and/or migratory movements. The classic example here is population fluctuations and migrations in lemmings.

Organisms are therefore (for the sake of simplicity) divided into r and K strategists, although of course there is a continuum between the two. R strategists tend to focus on rapid growth. They live only briefly and invest little in defense. Some plants even forgo cooperation with root fungi (mycorrhizae), such as cruciferous plants (Brassicaceae), since they do not live long enough to establish stable symbioses. It is our good fortune that cruciferous plants do not defend themselves with toxins and the like, because that is why we can eat them as cabbage, mustard, rutabaga, radish, cress, horseradish. R strategists benefit from disturbances of all kinds, since they usually have only short life cycles anyway, produce many seeds (if they are plants), are often relatively small, and reproduce quickly. Grasses are also good examples of r strategists; they benefit from disturbances, they practically live off the disturbance (i.e., grazing) by large herbivores. Without the herbivores, grasslands would become overgrown with shrubs, and the grasses would disappear. R strategists never reach K in their growth curves, as they are constantly kept well below their capacity limit by biotic or abiotic disturbances. R strategists are therefore dependent on disturbances and dynamics; without these, they would disappear. In Europe, without human intervention, grasslands would become forests (at least nowadays, since there is almost no naturally occurring megafauna left in Europe). R strategists are often asexual, i.e., parthenogenetic, because this accelerates their reproduction.

In contrast, the life cycles of K strategists are oriented more toward long-term survival. They invest heavily in growth and defense and reproduce relatively late in life. K strategists usually grow slowly toward the capacity limit K. Their densities then fluctuate (if at all) around the capacity limit. K strategists abhor disturbances, as in such cases the fast-growing and fast-reproducing species are favored. K strategists are more likely to be sexual. Examples of K strategists are sequoias or humans.

The r and K concept can also be illustrated by the survival rate of the individuals in a given population. K strategists (such as humans and some other mammals) have a manageable (i.e., small) number of offspring; they care for their young more or less intensively, and the survival rate of individuals is very high. Most individuals in these populations die only at a relatively old age. With r strategists, the exact opposite is the case. The number of offspring is extremely high (fungal spores, plant seeds, marine organisms such as sponges and corals that release thousands of gametes into the open water), the offspring usually have to fend for themselves, and parental care is rare. The death rate is already very high in early life stages, and only a few organisms survive to old age.

However, it is also the case that even within r and K strategists, some individuals are more or less r strategists, and some are more or less K strategists. Within most species, there is a gradual transition from more or less weakly expressed r to stronger r strategists; likewise with K strategists. Humans are a good example. Among us, there are also individuals who invest heavily in growth and resource accumulation (the conservatives) and some who reproduce early and invest less in resources (the book "Live Wild and Dangerously" by Osho probably reflects this attitude to life; Osho 2004). Even between the sexes (and certainly also within the sexes) there are differences. While women (in humans) and females (in the animal kingdom) tend to accumulate resources and reproduce later, men and males tend to invest in early reproduction. This often leads to conflicts between the sexes, which are all too familiar in both the animal kingdom and among humans.

We can therefore assume that within most species and populations, a gradient from r to K strategists exists. This means that some individuals can cope better with disturbances than others, and indeed, to a certain extent, even need disturbances. Ultimately, they depend on disturbances, as they have a competitive advantage in such situations. A disturbance does not necessarily have to be negative. It is only our judgment that sees disturbances and chaos as negative. Organisms that are more r-selected benefit from disturbances.

Now, the fact is: r-strategists are not better or worse than K-strategists (if one were inclined to make such a value judgment); they are simply different. This applies to animals and plants, and of course, all the more so to humans. Organisms have simply adapted in different ways to various abiotic and biotic environmental conditions. However, when it comes to humans, it is always difficult not to fall into value judgments. If someone reproduces early and makes little effort to secure resources for their children, he or she sometimes has a rather poor reputation. In our society, it is generally considered good and reasonable to work, to get an education, to save, not to have children too early (but also not too late), and then to enable them, in turn, to receive a good education, etc. The ecologist calls such an influence "stabilizing selection." This serves to criticize the extremes in society in order to stabilize the average behavior of the population and not allow too many outliers, so that the coherence and, in the long term, the stability of the community/society are preserved.

I would like to argue here that there is no objective justification for criticizing the exceptional positions within a society (unless they are outside the law). Everyone alive today looks back on a long line of ancestors who all successfully raised offspring. Therefore, there is also no justification for someone who has children early or late to be ostracized by society. Everyone should be able and allowed to live as he or she wishes. This view is by no means derived in any way from considerations of ecology or nature; it is my personal opinion. However, as an ecologist, I would claim that all our (very diverse) behaviors cannot be all that wrong; otherwise, they would not have left any descendants. So, first and foremost, we can all accept ourselves as we are.

What follows from these considerations? These reflections show (at least I hope so) that all our personal philosophy/ethics is always, to some extent, self-interested lobbying. If, as a person, I have many resources, I am automatically interested in a stable situation, since changes are unlikely to improve things for me. I am for stability, for peace, against chaos, and generally for maintaining the status quo; in short, I vote conservatively. The less I possess, the less interested I am in stability and order. Since I have nothing to lose anyway, chaos cannot harm me (things cannot get worse), and disruptions and crises do not bother me. In the extreme case, I am even explicitly in favor of disruptions, chaos, and revolutions, so that something finally changes (it can only get better); I even actively seek out and support such changes... however, it is difficult to openly advocate such a view... who dares to say that disruptions, chaos, and revolutions are positive... that seems outrageous and disreputable. But the strategy of grasses (which benefit greatly from disturbance) is neither better nor worse than that of trees (which generally do not like disturbances; except for fire-adapted trees, which only sprout after their seeds have experienced a forest fire). Overall, then, we should recognize: even if we desire peace and stability, this can be a form of self-interested lobbying—namely, because we are doing well. The better off we are, the more conservative we become. All these considerations should lead to humility regarding our value systems. Our values and norms are evidently dynamic; they change with our resource situation!

What is the purpose of this chapter? I wanted to show that the organisms of this earth line up along an r-K continuum (this is, of course, not the sole explanation for their occurrence; but it is an important aspect). This means that they are adapted to different degrees to various levels of disturbance. I also wanted to show that even within species, there are individuals who are more r- or more K-strategists. Some cope better with change and even benefit from it, while others tend to suffer from it. The philosophy and ethics that we as individuals derive from this are therefore dynamic. We should understand that many of our life concepts are shaped by our attitude toward disturbances and our resource availability. Thus, every ethic (e.g., no war, no revolution, no change, preserving everything as it is, etc.) is usually a form of self-interested lobbying. In effect, we represent our own interests and cloak them in an apparently objective ethic.

What now? First, we should view our own ethics and philosophy as more relative than before. We should critically examine the portion of our personal ethics that is primarily self-interested lobbying. Only then can we come to an understanding about our differing ethics, which arise from our life circumstances and may possibly lead to a higher-level consensus that takes into account the interests of both r-strategists and K-strategists. The K-strategists will always be interested in ensuring that, alongside themselves, many others also achieve prosperity (so that the status quo is not challenged; certainly no revolutions...), while the r-strategists have no objection to dynamism (and can thus drive progress). Both sides (and everyone in between) thus make a positive contribution to society, and this force should be harnessed to make society fit for the future and peaceful (although I am aware that this, of course, is once again self-interested lobbying; for obviously, I personally—at least at the moment—am more aligned with the K-strategists).

Reference

Osho (2004) Mut. Lebe wild und gefährlich. Ullstein, Berlin

Kin Selection—Are We All Racists? The Dark Side of Cooperation

The basic idea of Bill Hamilton's 'kin selection' theory is—as with so many brilliant ideas—quite simple. Bill Hamilton (1964a, b) argued that, for evolution, it is not solely important that the individual reproduces (although, of course, individual reproduction is an important factor in evolution), but that the reproduction of close relatives can also indirectly contribute to the success of one's own genes. This is why the theory was also called the theory of kin selection. It states, for example, that it is equivalent whether I have offspring myself or two of my siblings do (with whom I am, on average, 50% related). In both cases, my reproductive success would be identical. In this context, "my" refers to the genes, which—on average—are present in me in exactly the same proportions as in my two siblings combined. This also explains why most parents would sacrifice themselves for their children, as this ensures their (and their own) reproductive success.

It is thus clear that the 'interest' of the genes is at the center of evolution. However, in evolution, it is never individual genes that live and die, but only individuals. On which level (genes, individuals, groups, populations, species) evolution ultimately operates is an intensely and controversially discussed topic that we cannot examine in detail here. For our topic, however, it is of little importance.

This perspective of genes (or alleles) that "want" to make it into the next generation (in whatever way) was later popularized by Richard Dawkins in his book "The Selfish Gene" (Dawkins 1976). This book was later often criticized as social Darwinist, which, however, is based on a misunderstanding. Dawkins has been and continues to be criticized for supporting a view of evolution in which genes are only selfish and in which the survivor is the one who is the least cooperative (just as a reminder: by definition, the successful genes are those that have made it into the next generation—by whatever means). The misunderstanding arose from the term "selfish gene." This reminded many people of "survival of the fittest"

M. Maraun, *Everything is Ecology*, https://doi.org/10.1007/978-3-658-50207-2_8

and thus automatically of a nature as described by Alfred Tennyson in his 1850 poem "In Memoriam A.H.H.": "Nature, red in tooth and claw," which stands for the cruelty of nature. But that is not what Richard Dawkins meant. To make it into the next generation, it is just as possible to cooperate as to be selfish. Ruthlessness and selfishness may be one way to reach the next generation; cooperation and symbiosis (i.e., collaboration between different species) are another way to reach the next generation. This cooperation is not meant as altruism, as I have already explained in the Chap. "Tragedy of the commons and altruism." Perhaps Dawkins' book would have been received differently if it had had a different title. Dawkins himself later said that he could have called his book "The Cooperative Gene"; it would have made no difference in content. Many paths can lead to the success of a gene. A clever strategy can also be to simply insert oneself into the genome without any function, as many so-called "transposable elements" do, which we then refer to as "junk DNA."

Building on Hamilton's ideas, Dawkins was able to explain a large part of the so-called altruism in the animal kingdom with the hypothesis of 'kin selection,' which had been an unsolved mystery for generations of biologists. Why should we cooperate when it seems so much more plausible to be selfish? Selflessness and cooperation have been a problem for biologists for centuries. Even Charles Darwin doubted his own theory and wrote, "altruism presents a special difficulty, which at first appears to me insuperable, and actually fatal to my whole theory. [...] How can natural selection, which acts solely by and for the benefit of each individual, explain the behaviour of social insects which seem often to sacrifice their own welfare for the benefit of the community?" (Darwin 1871). Robert May added many years later: "The most important unanswered question in evolutionary biology, and more generally in the social sciences, is how cooperative behaviour evolved and can be maintained in human or other animal groups and societies." (May 2006).

Does the 'kin selection' theory really explain all apparently altruistic behavior, and can it ultimately be reclassified as selfish? In the Chap. "Tragedy of the commons and altruism" I have already briefly addressed altruism; however, since the topic is of such outstanding importance, I would like to briefly address it again here. There are many examples of seemingly altruistic behavior: meerkats risk their lives when they keep watch for predators (mostly birds of prey); vampire bats share resources (blood) with related and unrelated bats that are currently experiencing food shortages; baboons groom conspecifics; birds sometimes help conspecifics raise their young; and the gut bacterium *Escherichia coli* even commits suicide when infected by a bacteriophage that could also infect other *E. coli*. The behavior of some individuals among eusocial animals, for example various species of bees, wasps, ants, and termites, and even in one vertebrate species (naked mole rats), who completely forgo their own reproduction in favor of a queen (in termites, in favor of king and queen), appears even more altruistic. To put it briefly: all these behaviors are various forms of kin selection, i.e., support of relatives, or of reciprocal altruism, which could simply be called trade (in the sense of: you help me now; I'll help you later). Not a single case is true altruism in the sense that one actor loses something while another benefits. Meerkats mainly keep watch for predators to protect members of their own family; vampire bats give blood when other individuals are hungry, but remember to whom they have given and later demand blood in return when they themselves are short of food (reciprocal altruism); baboons groom each other so that both sides benefit in the end; and birds mainly help closely related conspecifics, so this behavior is also selfish in the sense of 'kin selection.' They also usually help only as long as they themselves do not have an opportunity to breed. The bacterium *E. coli* is a special case. Suicide seems to be the ultimate case of altruism. However, when *E. coli* eliminates itself (because it is infected by a bacteriophage or similar, which also threatens to infect its clones), there seems to be no explanation in terms of selfishness. However, the bacteria sacrifice themselves for their "relatives" mainly when they are truly genetically clonal and thus 100% genetically identical. This behavior has nothing to do with altruism or self-sacrifice and is ultimately actually selfish, because if I protect my clones (which are 100% identical to me), I am also protecting myself. The same applies to the cells of an organism, which—as in humans—forego their own reproduction (except, of course, for the germline cells). They do this only because they are genetically 100% identical to the germline cells, and thus ultimately serve their own interests. The fact that such a system does not always work perfectly is shown by the existence of cancer; in this case, cells that do not belong to the germline suddenly divide, which ultimately harms the entire organism.

So it really seems that there is no true altruism. But what about the even harder-to-understand case of forgoing one's own reproduction in eusocial animals such as bees, wasps, and ants (let's leave termites aside for now, as their colonies are organized differently from those of the hymenopterans—that is, bees, wasps, and ants)? The individuals of these species are not clonally identical; so this explanation does not apply when it comes to explaining this renunciation of reproduction. Through the mating of the queen (with a diploid (double) set of chromosomes) with a male drone (which has a haploid (single) set of chromosomes), there is a high degree of relatedness among the workers (who arise from fertilized eggs of the queen). They are (at least theoretically) 75% related to each other. For comparison: we are, on average, 50% related to our siblings.

The common 'kin selection' theory now states that these siblings (the workers) cooperate so well and do not produce their own offspring because they are so closely (75%) related. The 75% results from the average of the DNA that comes from the mother (50%) and the DNA that comes from the father (100%). In this way, they help the group (for example, the entire beehive) to be successful, and also support the queen in raising the offspring.

A beautiful theory; however, it has two flaws. First of all, the workers in large colonies of bees, wasps, and ants are not as closely related to each other as the theory suggests. Why is that? During the nuptial flight, the queen (at least in species that live in large colonies) usually mates with several drones (males), so the relatedness of the workers in the hive ultimately averages around 30%, which is significantly lower than if she had mated with a single drone (in that case, as mentioned above, the workers would be 75% related) and even significantly lower than the relatedness we have with our own siblings (50% on average—as mentioned above). So, there is actually no reason to cooperate so closely with the other workers.

Secondly, the workers (if they were to lay eggs) would be much more closely related to their own sons than to those of the queen, since the drones arise from unfertilized eggs laid by the workers (the workers have not undergone a nuptial flight for fertilization and can therefore only lay haploid eggs). The eggs of a worker that could develop into drones would be 100% derived from that worker; whereas the queen's offspring (regardless of whether she fertilizes these eggs (producing female workers) or not (males = drones)) would in any case be less related to the worker.

Can the workers of eusocial hymenopterans (bees, wasps, ants, etc.) even lay their own unfertilized eggs, which then develop into drones that, so to speak, become the sperm of the colony (since they will fertilize the next queen)? Yes, some workers have functional ovaries and do indeed lay their own eggs, which can also develop into drones. However, in most large colonies of hymenopterans, so-called "policing" occurs, meaning that other workers monitor whether some workers are secretly laying eggs that could develop into drones. These eggs are then either eaten or at least destroyed by the other workers. This gives you an idea of how "cooperative" the sisters in the beehive really are with each other. They do not allow each other any eggs or offspring, since (due to the queen's frequent mating during the nuptial flight) they are only so distantly (~30%) related to each other—as explained above. The queen's trick to gain control in, for example, a bee colony, is therefore to mate with many males, in order to keep the relatedness of the workers in the hive low, so that they do not band together against the queen and/or lay their own eggs. Is this now altruism or cooperative behavior? Certainly not. The evolution of social behavior in bees and other hymenopterans can thus be imagined as follows: In small colonies at the beginning of the evolution of eusociality (eusociality = the renunciation of reproduction by some individuals in a population), the queen mated with only one male; she then laid a few eggs; the relatedness of the workers was now truly high (75%; see above) and there was no policing (due to the good cooperation among the workers). The queen herself had to prevent the reproduction of the workers by eliminating the workers' eggs herself (direct control by the queen). In the slowly—over evolutionary timescales—growing colonies, she could no longer maintain this control herself. She therefore invented (evolved) the trick of mating with several drones during the nuptial flight, so that the relatedness of the workers decreased. This led to "policing," that is, the elimination of workers' eggs by other workers. As a result, only a few workers produced eggs. However, to this day, a small proportion of workers have retained the ability to lay their own eggs, which are sometimes, but not always, destroyed or eaten by other workers. Thus, even in large colonies, the queen controls the reproduction of the workers; but in this case, indirectly (indirect control by the queen). So, are bees and other hymenopterans an example of a perfectly cooperating superorganism, as such colonies are sometimes called (Hölldobler and Wilson 2009), and thus an example of altruism and harmony? By no means. Cooperation in the beehive is enforced by the queen and should therefore by no means be regarded as an example of altruism by the non-reproducing workers. So, let us reiterate: There is no altruism in the animal kingdom, since in evolution, interactions cannot prevail that are purely negative for one participant and purely positive for the other. But what about us humans? Is there also no altruism among us; or are we an exception in the realm of organisms?

In any case, we are very social. We help others in need; and we generally work well in groups, even when we are not related to the members. We recognize our group members by physical features or (more often) by rituals. Most religious communities also have clothing or other features that make group membership easily recognizable. We are not eusocial (i.e., only very few individuals voluntarily forgo reproduction), but we live together in groups with large numbers of individuals relatively peacefully. Such large groups would lead to very significant conflicts in other primates. We trade with other groups; we organize ourselves into overarching groups (cities, countries, EU, NATO) to which we cede part of our autonomy.

We are even so social that we sanction group members who do not act socially. One of the most famous examples of this is an experiment in which a test subject, for example, receives €100 (from an experimenter) and can give any amount (or nothing at all) to another person. This other person can reject the deal; then neither of them gets anything. Now, you would think that the person who receives a portion of the €100 would accept anything (except zero euros); after all, the alternative is that they get nothing. In fact, however, most participants only accept the deal if they would receive about a third of the €100. This means that they would rather harm themselves (i.e., let the deal fall through for €0 to ~€30, so that neither gets anything) than let the owner of the €100 get away with their antisocial behavior. This experiment very clearly shows that we are very social and that we are even willing to accept disadvantages ourselves in order to punish antisocial behavior. So, could forgoing the €0 to €30 be seen as a kind of altruism? By punishing antisocial behavior, we can sometimes harm ourselves. Such behavior is sometimes called "third party punishment." We can already anticipate that this behavior is also not altruism, since it stabilizes the group and its norms (from which I ultimately benefit myself).

With that, we now leave the world of cooperation and turn to the dark side of cooperation; here it is about control, deception, and discrimination. And by definition: If we discriminate positively (and almost all of us love our own children and the children of our siblings more than those of our neighbors; = 'kin selection'), we automatically do so negatively (against someone else), otherwise positive discrimination would be pointless. The group that supports each other positively does so in contrast to another group. Otherwise, it would not be a group; otherwise, it would be 'everyone.' Just as a reminder: that was exactly the reason why the United Nations do not work; they have no group against which they can define themselves.

Since we are talking about discrimination (which at first simply means a separation into an in-group and an out-group), we should briefly touch on the topic of racism; it cannot be avoided anyway. What is racism, anyway? The question itself is already nonsense, because biology shows us that there are no races. In biology, race is not defined. One can define a species (all individuals that are capable of interbreeding), but even that is not always easy. What about, for example, "different" plant species that can hybridize; where is the species boundary there? What about asexual 'species,' which get by without males? In that case, each individual would be a closed reproductive 'community,' since each female is capable of producing offspring on her own. What about fossils, which by definition cannot produce fertile offspring with anyone at all? The species in time is therefore poorly defined. Moreover, species can split into new species. In this transitional phase, the species boundary is hard to determine. Almost all species concepts ignore the dynamics of species boundaries. If defining a species is already so difficult, how much harder is it to define a race, a subspecies, or something similar? The genetic differences and similarities among individuals of a species are therefore fluid.

Within the species Homo sapiens (*Homo sapiens*), there is thus no way to define a race. Therefore, there is also no racism (in the strict sense of the word), and we therefore cannot prohibit discrimination based on race (for example, in the German Basic Law). It therefore seems reasonable to remove the term "race" from the Basic Law. This is also supported by many political parties. If the term race is thus removed, it only seems logical to also remove the term racism.

If we now remove the concepts of "race" and "racism" from our repertoire, this does not mean that discrimination does not exist. As I have tried to show above with the idea of 'kin selection', (positive) discrimination (the support of relatives; sometimes also recognized negatively as nepotism) is an essential component of living beings. Since there is positive recognition and support, this automatically means that there is also negative discrimination; the two are inextricably linked. I think it is important that we acknowledge the following: We all discriminate positively (towards our in-group; relatives, etc.; kin selection), and by doing so, we automatically discriminate against the out-group. What does this mean? First of all, nothing, except that we should acknowledge this problem. And above all, it does not mean that we have to take a resigned attitude towards this issue (that's just how it is... we all discriminate); rather, as human beings, we have a choice in how we want to deal with this topic. We really do have a choice! We have chosen against slavery, against child labor, against the oppression of women, and we can also decide how we, as humans, want to deal with the topic of 'kin selection'. Because even if we acknowledge that there are conflicts and discrimination, it is up to us to establish rules to minimize such conflicts and problems. We are free to write the history of humanity. As I already wrote in the prologue, we can make a difference. But we should not forget that we are living beings on this earth, who have arisen in the course of evolution and are therefore not completely free from the rules of ecology and evolution.

The 'kin selection' theory also leads quite compellingly to the conclusion that we will always favor relatives (financially or with jobs or similar), at least if we are not prevented from doing so. And since we do not want to establish royal or dictatorial ruling dynasties (because, in general, these have not led to the best rulers in history, and because this is clearly not in line with the idea of democracy), we must do everything to prevent nepotism. "Transparency International," an organization that combats corruption and nepotism in politics and business, should therefore be supported as much as possible.

But let us return once more to our finding. Can we solve the problem of discrimination that arises from the theory of 'kin selection'? In my opinion, it first means that we should all accept the element of discrimination within ourselves. This does not mean that we should act on it (!); but we may acknowledge it as a feeling. We simply love our children and relatives more than the 'others.' This already allows us to understand each other a little better. It also gives us more empathy for the 'others' who feel similarly. And empathy is always a good thing. In general, we would be less arrogant towards the 'other' and could show a bit more humility. That, too, seems positive to me. But the question is, how do we deal with this in everyday life? Of course, the goal is to reduce discrimination and to reduce or even prevent all its negative consequences. Just to avoid any misunderstandings: Of course, we want to prevent exclusion, abolish apartheid, appreciate one another, see diversity and variety as positive, and let everyone live as he or she wishes. And to be even clearer: I support the liberal ideas of equality and justice; and I support all green ideas that seek to minimize and ultimately end discrimination and exclusion. I would like to explicitly block any appropriation from the right here. But I think we should also acknowledge the discriminatory part within ourselves. Only then can we honestly address the issue of discrimination, transcend the problem, and truly leave it behind us.

References

Darwin C (1871) The descent of man. John Murray, London

Dawkins R (1976) The selfish gene. Oxford University Press, Oxford

Hamilton WD (1964a) The genetical evolution of social behaviour I. J Theor Biol 7, 1–16–39

Hamilton WD (1964) The genetical evolution of social behaviour II. J Theor Biol 7:17–52

Hölldobler B, Wilson EO (2009) Der Superorganismus: Der Erfolg von Ameisen, Bienen, Wespen und Termiten. Springer

May RM (2006) Address of the President, Lord May of Oxford OM AC FRS, given at the Anniversary Meeting on 30 November 2005: threats to tomorrow's world. Notes Rec R Soc 60:109–130

Epilogue and Outlook

Why this book? I wanted to show that, on the one hand, as humans we are quite ordinary organisms. We interact with our environment, we acquire resources, our population grows to its (admittedly flexible) limit K (K = capacity), we compete, we cooperate (though most of the time we cooperate in order to outcompete others). And this will always be the case as long as living beings exist on Earth. The idea of the end of history (Fukuyama 1992) is therefore absurd; history and evolution always continue. Currently, we see the problems of massive global resource acquisition and overuse, environmental pollution, and global warming. Everywhere we see the tragedy of the commons; everyone wants environmental protection, healthy food, wind turbines, clean water, but no one wants to pay for it, and it should be done and paid for by others; and no one wants wind turbines in their own backyard. This will never change as long as there are organisms. This is not a plea for agony, cynicism, or fatalism. For we are capable of great progress and of humanism. There is (in Europe) good healthcare for almost everyone, child labor is banned, slavery has been abolished, we no longer burn women as witches; and there is less hunger in the world, fewer wars than before, we have significantly increased K (the available resource), we are 10 billion people on Earth (which is, at first, a success story), we have functioning sanitation systems, and we are slowly beginning to understand that the Earth is finite, and that despite all progress we cannot consume more than is available. There are, therefore, limits to growth (the "Club of Rome" knew this decades ago; Meadows et al. 1972). Moreover, technological progress (e.g., desulfurization plants during the era of acid rain) is capable of solving some problems (the ozone hole is no longer a major issue, polluted rivers are partially recovering; the reintroduction of wolves and beavers is a success for conservation; overpopulation is currently no longer an issue; acid rain no longer destroys our forests; we already generate much energy from renewable sources; and perhaps we can also technically fix

M. Maraun, *Everything is Ecology*, https://doi.org/10.1007/978-3-658-50207-2_9

CO_2 (although plants are better at this)). And to be absolutely clear: I am absolutely optimistic and consider it irresponsible to write books that predict the end of the world every few years (e.g., "So let us plant an apple tree" by Hoimar von Ditfurth in the 1980s; von Ditfurth 1985). Of course, there is much to be done, but it is annoying when it is always five minutes to midnight.

After a long paragraph about the one side, now comes the other. On the other hand, we are also very atypical animals, we live in very large groups, we are seekers of meaning, we thirst for ideas and ideologies, for meaning and for the creation of meaning. Since we are capable of self-reflection, this will always remain so. Who fulfills this longing? Usually, we seek this meaning in religions, in sects, in ideologies (and in everything that ends with -ism, e.g., communism, capitalism, socialism, liberalism), in esotericism, in astrology, Bach flower remedies, homeopathy, in the supernatural, and so on. All these—partly irrational—things will, in the medium and long term, become less and less prevalent; religions are slowly disappearing or splintering into sects (Mormons, Hutterites, Amish, Scientology, ...); Enlightenment has deprived them of any foundation—even if this will still take centuries. Ideologies (and the -isms) will become increasingly irrelevant. So what will replace all these irrational ideologies in the future? Where is humanity heading in the best case? What can unite us (beyond egoism and competition)?

Should the irrational reasons for group cohesion fall away, we will need new visions that inspire us as humans and that can also be rationally shared by everyone (there is nothing worse than Helmut Schmidt's saying: "If you have visions, you should see a doctor"; on the contrary: "It is precisely our visions that make us human"). These visions, which should have an almost sacred character, will include genuine values such as freedom, justice, democracy, human rights, a free press, independent courts, protection of minorities, and living together in peace. The importance of these values (which are worth fighting for) is, in my opinion, underestimated and taken for granted in our Western societies. We in the West have grown up with them, we have received all this as a gift. How precarious, endangered, and fragile such values are can be seen in people like Donald Trump and Jair Bolsonaro, for whom egoism and the preservation of power take precedence over all values. However, given the egoism I have already mentioned and discussed above, this should at least not surprise us. All the more important, then, is to prevent and stop such behavior. The invasion of Ukraine by Russia also shows how little peace, democracy, freedom, and human rights can be taken for granted.

So what is to be done? Elections (the core of democracy) must be able to be thoroughly scrutinized (it cannot be that there are already first results one minute after the polls close); all ballots should be analog (digital election results are hardly verifiable), there should be no polls for a month before elections (because polls can influence elections), the confusion of first and second votes (in Germany) should be abolished (only one third of the population knows the meaning of both votes), the nonsense of overhang and compensatory mandates and the resulting constant enlargement of the German Bundestag must stop, and every vote must count exactly the same (so there should only be one vote in elections); with the latter, arbitrary redistricting (as in the USA) would also be pointless. There should be a discussion about whether proportional representation or majority voting is "better." At what threshold should a party be allowed to have seats in parliament? An election must become a significant event; fair, free, and verifiable by anyone afterwards.

Humanism must also replace religions and all other forms of superstition. No idea (except, e.g., democracy, freedom, human rights) must ever be more important than a human life. People should respect each other and recognize their diversity as valuable.

None of this will come to us easily. It will be a long and arduous path, for biology tells us: we are egoists, nepotists, and we discriminate. That is why organizations like "Transparency International" are of immense importance, because we have no self-control; almost everyone is susceptible to bribery and temptation; therefore, when it comes to self-control, we must, like Odysseus, tie ourselves to the mast to avoid being devoured by the sirens of self-interest and egoism. Precisely because biology has made us aware of our weaknesses in this regard (see above).

The cooperation of our liberal, free societies should therefore be guided by how autocrats in Russia, China, Iran, and North Korea (as of spring 2025; this can change quickly) behave, because nothing pleases these countries more than when we forget that in our societies, the individual has intrinsic value. Dictatorships and autocracies want us to sink to their level; and that is the worst thing that can happen to us; they want us to use violence as well and betray our ideals.

Therefore, economic cooperation with dictatorships and autocrats should be reduced, and cooperation with democracies, such as India and Brazil, should be promoted. An independent commission (then the UN would finally have a function) could assess the state of democracy, free press, and human rights in the countries of the world (of course, this commission would be very, very susceptible to corruption). This concerns not only economic cooperation, but also cooperation in the fields of science and culture.

In the end, this also helps people in countries with dictatorships. Germany is one of the best examples of how incapable a country can be of winning democracy and freedom for itself (at least East Germans managed it in 1989), and how, after many decades, the idea of a free, liberal democracy can take root even in a country (Germany) that, in the 20th century, covered the world with fascism, war, destruction, and the persecution of Jews. So there is hope.

All this now has little to do with ecology (but is not completely detached from it, as I have tried to show in this book), but rather is my hope for how things could go in the best case. And I am confident that we humans can achieve this. In our own best interest, and in that of our children and grandchildren.

References

Fukuyama F (1992) Das Ende der Geschichte. Kindler, München.

Meadows D, Meadows D, Zahn E, Miling P (1972) Die Grenzen des Wachstums. Bericht des Club of Rome zur Lage der Menschheit

Von Ditfurth H (1985) So lasst uns denn ein Apfelbäumchen pflanzen: Es ist soweit. Rasch und Röhring, Hamburg

FSC
www.fsc.org
MIX
Papier | Fördert
gute Waldnutzung
FSC® C083411

Kolibri 360 GmbH
Ferdinand-Jühlke-Straße 7
99095 Erfurt, Deutschland
produktsicherheit@kolibri360.de